FOUNDATIONS OF HIGH PERFORMANCE POLYMERS

Properties, Performance, and Applications

FOUNDATIONS OF HIGH PERFORMANCE POLYMERS

Properties, Performance, and Applications

Edited by

Abbas Hamrang, PhD, and Bob A. Howell, PhD

Gennady E. Zaikov, DSc, and A. K. Haghi, PhD
Reviewers and Advisory Board Members

Apple Academic Press

TORONTO   NEW JERSEY

Apple Academic Press Inc. | Apple Academic Press Inc.
3333 Mistwell Crescent | 9 Spinnaker Way
Oakville, ON L6L 0A2 | Waretown, NJ 08758
Canada | USA

©2014 by Apple Academic Press, Inc.

First issued in paperback 2021

Exclusive worldwide distribution by CRC Press, a member of Taylor & Francis Group
No claim to original U.S. Government works

ISBN 13: 978-1-77463-281-9 (pbk)
ISBN 13: 978-1-926895-52-9 (hbk)

Library of Congress Control Number: 2013945481

Library and Archives Canada Cataloguing in Publication

Foundations of high performance polymers: properties, performance, and applications/ edited by Prof. A. Hamrang, and Bob A. Howell, Gennady E. Zaikov, DSc, and A.K. Haghi, PhD, reviewers and advisory board members.

Includes bibliographical references and index.
ISBN 978-1-926895-52-9
1. Polymers. 2. Polymeric composites. 3. Polymers--Industrial applications.
4. Nanofibers. I. Hamrang, A. (Abbas), author, editor of compilation

TA455.P58F69 2013   620.1'92   C2013-904994-0

Apple Academic Press also publishes its books in a variety of electronic formats. Some content that appears in print may not be available in electronic format. For information about Apple Academic Press products, visit our website at **www.appleacademicpress.com** and the CRC Press website at **www.crcpress.com**

ABOUT THE EDITORS

Abbas Hamrang, PhD

Abbas Hamrang, PhD, is currently a Senior Polymer Consultant and Associate Editor and member of the academic board of international journals. He was previously an international Senior Lecturer-Professor of polymer science and technology, Deputy Vice-Chancellor (R & D), R & D Co-ordinator, Manufacturing Manager, Manufacturing Consultant, Science & Technology Advisor (all at international level) in both the academic and industrial sectors. Abbas Hamrang, PhD, is currently a senior polymer consultant and an associate editor and member of the academic board of international journals. He was previously an international Senior Lecturer-Professor of polymer science and technology, Deputy Vice-Chancellor (R & D), R & D Co-ordinator, Manufacturing Manager, Manufacturing Consultant, Science & Technology Advisor (all at international level) in both the academic and industrial sectors. His specialist area is in polymer degradation and stabilization.

Bob A. Howell, PhD

Bob A. Howell, PhD, is a professor in the Department of Chemistry at Central Michigan University in Mount Pleasant, Michigan. He received his PhD in physical organic chemistry from Ohio University in 1971. His research interests include flame retardants for polymeric materials, new polymeric fuel-cell membranes, polymerization techniques, thermal methods of analysis, polymer-supported organoplatinum antitumor agents, barrier plastic packaging, bioplastics, and polymers from renewable sources.

REVIEWERS AND ADVISORY BOARD MEMBERS

Gennady E. Zaikov, DSc

Gennady E. Zaikov, DSc, is Head of the Polymer Division at the N. M. Emanuel Institute of Biochemical Physics, Russian Academy of Sciences, Moscow, Russia, and professor at Moscow State Academy of Fine Chemical Technology, Russia, as well as professor at Kazan National Research Technological University, Kazan, Russia. He is also a prolific author, researcher, and lecturer. He has received several awards for his work, including the Russian Federation Scholarship for Outstanding Scientists. He has been a member of many professional organizations and on the editorial boards of many international science journals.

A. K. Haghi, PhD

A. K. Haghi, PhD, holds a BSc in urban and environmental engineering from the University of North Carolina (USA); a MSc in mechanical engineering from North Carolina A&T State University (USA); a DEA in applied mechanics, acoustics and materials from Université de Technologie de Compiègne (France); and a PhD in engineering sciences from Université de Franche-Comté (France). He is the author and editor of 65 books as well as 1000 published papers in various journals and conference proceedings. Dr. Haghi has received several grants, consulted for a number of major corporations, and is a frequent speaker to national and international audiences. Since 1983, he served as a professor at several universities. He is currently Editor-in-Chief of the *International Journal of Chemoinformatics and Chemical Engineering* and *Polymers Research Journal* and on the editorial boards of many international journals. He is also a faculty member of University of Guilan (Iran) and a member of the Canadian Research and Development Center of Sciences and Cultures (CRDCSC), Montreal, Quebec, Canada.

CONTENTS

List of Contributors.. *xi*

List of Abbreviations ....*xv*

Preface .... *xvii*

Section 1: Foundations of High Performance Polymers

1. **Commercially Produced and Modified Starches: Part I**........................ 1

   A. I. Sergeev, N. G. Shilkina, L. A. Wasserman, and H. Staroszczyk

2. **Commercially Produced and Modified Starches** 13

   S. Z. Rogovina, K. V. Aleksanyan, and E. V. Prut

3. **Production and Application of Polymer Nanocomposites: Part I** 35

   G. V. Kozlov, K. S. Dibirova, G. M. Magomedov, Bob A. Howell,
   and G. E. Zaikov

4. **Production and Application of Polymer Nanocomposites: Part II** 43

   G. V. Kozlov, Z. M. Zhirikova, V. Z. Aloev, Bob A. Howell, and
   G. E Zaikov

5. **Computation Modeling of Nanocomposites Action on the
   Different Media**.. 53

   V. I. Kodolov, N. V. Khokhriakov, V. V. Trineeva, M. A. Chashkin,
   L. F. Akhmetshina, Yu. V. Pershin, and Ya. A. Polyotov

6. **Application of Genetic Engineering: Part I** 113

   A. S. Korovashkina, S. V. Kvach, and A. I. Zinchenko

7. **Application of Genetic Engineering: Part II**....................................... 125

   A. S. Korovashkina, S. V. Kvach, M. V. Kvach, A. N. Rymko,
   and A. I. Zinchenko

Section 2: Production of Nanofibers

8. **Nanopolymers: Fundamentals and Basic Concepts** 139

   A. K. Haghi and A. Hamrang

9. **Nanomaterials: Dos and Don'ts of Production Process in Laboratory**............ 195

   A. K. Haghi and A. Hamrang

10. **Green Nanofibers**...271

   A. K. Haghi, V. Mottaghitalab, M. Hasanzadeh, and
   B. H. Moghadam

   Index...383

LIST OF CONTRIBUTORS

L. F. Akhmetshina
OJSC "Izhevsk Electromechanical Plant—Kupol"

K. V. Aleksanyan
Semenov Institute of Chemical Physics, Russian Academy of Sciences, ul. Kosygina 4, Moscow 119991, Russia

V. Z. Aloev
Kabardino-Balkarian State Agricultural Academy, Tarchokov Str. 1a, Nal'chik 360030, Russian Federation, E-mail: I_dolbin@mail.ru

M. A. Chashkin
OJSC "Izhevsk Electromechanical Plant – Kupol"

K. S. Dibirova
Dagestan State Pedagogical University, Yaragskii Str. 57, Makhachkala 367003, Russian Federation, E-mail: I_dolbin@mail.ru

A.K. Haghi
University of Guilan, Rasht, Iran

A. Hamrang
Senior Polymer consultant, Manchester, UK

M. Hasanzadeh
University of Guilan, Rasht, Iran

Bob A. Howell
Central Michigan University, Mount Pleasant, MI, USA, E-mail: Bob.A.Howell@cmich.edu

N.V. Khokhriakov
Basic Research—High Educational Centre of Chemical Physics and Mesoscopy, Udmurt Scientific Centre, Ural Division, Russian Academy of Sciences, Izhevsk Izhevsk State Agricultural Academy

V. I. Kodolov
M.T. Kalashnikov Izhevsk State Technical University
Basic Research—High Educational Centre of Chemical Physics and Mesoscopy, Udmurt Scientific Centre, Ural Division, Russian Academy of Sciences, Izhevsk

A. S. Korovashkina
Institute of Microbiology, National Academy of Sciences, Kuprevich Str. 2, Minsk 220141, Belarus

G. V. Kozlov
Dagestan State Pedagogical University, Yaragskii Str. 57, Makhachkala 367003, Russian Federation
Kabardino-Balkarian State Agricultural Academy, Tarchokov Str. 1a, Nal'chik 360030, Russian Federation, E-mail: I_dolbin@mail.ru

M. V. Kvach
Institute of Microbiology, National Academy of Sciences, 220141, Kuprevich Str. 2, Minsk 220141, Belarus

S. V. Kvach
Institute of Microbiology, National Academy of Sciences, Kuprevich Str. 2, Minsk 220141, Belarus

B. H. Moghadam
University of Guilan, Rasht, Iran

G. M. Magomedov
Dagestan State Pedagogical University, Yaragskii Str. 57, Makhachkala 367003, Russian Federation, E-mail: I_dolbin@mail.ru

V. Mottaghitalab
University of Guilan, Rasht, Iran

Yu.V. Pershin
M.T. Kalashnikov Izhevsk State Technical University
Basic Research—High Educational Centre of Chemical Physics and Mesoscopy, Udmurt Scientific Centre, Ural Division, Russian Academy of Sciences, Izhevsk

Ya. A. Polyotov
M.T. Kalashnikov Izhevsk State Technical University
Basic Research—High Educational Centre of Chemical Physics and Mesoscopy, Udmurt Scientific Centre, Ural Division, Russian Academy of Sciences, Izhevsk

E. V. Prut
Semenov Institute of Chemical Physics, Russian Academy of Sciences, ul. Kosygina 4, Moscow 119991, Russia

S. Z. Rogovina
Semenov Institute of Chemical Physics, Russian Academy of Sciences, ul. Kosygina 4, Moscow 119991, Russia

A. N. Rymko
Institute of Microbiology, National Academy of Sciences, Kuprevich Str. 2, Minsk 220141, Belarus

A. I. Sergeev
Institute of Chemical Physics RAS, Kosygin's Str. 4, Moscow, Russia, E-mail: nismpa@mail.ru

N. G. Shilkina
Institute of Chemical Physics RAS, Kosygin's Str. 4, Moscow, Russia
E-mail: nismpa@mail.ru

H. Staroszczyk
Gdansk University of Technology, ul. G.Narutowicza 11/12, 80-233 Gdańsk, Poland, E-mail: hanna.staroszczyk@pg.gda.pl

V. V. Trineeva
Basic Research—High Educational Centre of Chemical Physics and Mesoscopy, Udmurt Scientific Centre, Ural Division, Russian Academy of Sciences, Izhevsk
Institute of Mechanics, Ural Division, Russian Academy of Sciences, Izhevsk

L. A. Wasserman
Institute of Biochemical Physics RAS, Kosygin's Str. 4, Moscow, Russia, E-mail: lwasserma@mail.ru

G. E. Zaikov
N.M. Emanuel Institute of Biochemical Physics, Russian Academy of Sciences, Kosygin Str. 4, Moscow 119334, Russian Federation, E-mail: Chembio@sky.chph.ras.ru

Z. M. Zhirikova
Kabardino-Balkarian State Agricultural Academy, Tarchokov Str. 1a, Nal'chik 360030, Russian Federation, E-mail: I_dolbin@mail.ru

A. I. Zinchenko
Institute of Microbiology, National Academy of Sciences, Kuprevich Str. 2, Minsk 220141, Belarus, E-mail: zinch@mbio.bas-net.by

LIST OF ABBREVIATIONS

AcAc	Acetiylacetone
ACA	Amino caproic acid
AChE	Acetylcholinesterase
AFD	Average fiber diameter
ANN	Artificial neural network
ANOVA	Analysis of variance
APPh	Ammonium polyphosphate
BSSE	Basis superpositional error
CA	Contact angle
$CaCO_3$	Calcium carbonate
CB	Carbon black
CCD	Central composite design
c-di-GMP	3′, 5′-Cyclic diguanylate
CDS	Clinical Diagnostic Station
CF	Carbon fibers
CHT	Chitosan
CMC	Cell membrane complex
CNT	Carbon nanotubes
CVD	Chemical vapor deposition
DAcA	Diacetonealcohol
DAQ	Data acquisition
DD	Deacetylation degree
DGC	Diguanylate cyclase
DLS	Dynamic light scattering
DMF	Dimethylformamide
DSC	Differential scanning calorimetry
DWNT	Double-walled carbon nanotube
ECG	Electrocardiogram
ER	Epoxy resin
FS	Fine suspensions
GTP	Guanosine-5′-triphosphate
HDPE	High density polyethylene
HP	High pass
ICD	Implantable Cardioverter Defibrillator
LBM	Lattice Boltzmann method

LMC	Low-molecular compound
LMCS	Low molecular weight chitosan
LP	Low pass
MFD	Mean fiber diameter
MWNTs	Multi-walled carbon nanotubes
NS	Nanostructures
ODE	Ordinary differential equation
PAN	Polyacrylonitrile
PCL	Polycaprolactone
PCM	Polymeric composite materials
PEO	Polyethylene oxide
PEPA	Polyethylene polyamine
PP	Polypropylene
RI	Relative importance
RMSE	Root mean square errors
RSM	Response surface methodology
SDS	Sodium dodecyl sulfate
SDS-PAGE	SDS polyacrylamide gel electrophoresis
SEM	Scanning electron microscope
SI units	System International d'unites
StdFD	Standard deviation of fiber diameter
SWNTs	Single-walled carbon nanotubes
TiO_2	Titanium dioxide
UCM	Upper-convected Maxwell
WSN	Wireless Sensor Networks
ZPVE	Zero-point vibration energy

PREFACE

Polymers have a significant part in human existence. They have a role in every aspect of modern life, such as health care, food, information technology, transportation, energy industries, etc. The speed of developments within the polymer sector is phenomenal and at same time crucial to meet the demands of life today and in the future. Specific applications for polymers range from adhesives, coatings, painting, foams, and packaging to structural materials, composites, textiles, electronic and optical devices, biomaterials, and many other uses in industries and daily life. Polymers are the basis of natural and synthetic materials. They are macromolecules and in nature are the raw material for proteins and nucleic acids, which are essential for the human body.

Cellulose, wool, natural and synthetic rubber, and plastics are well-known examples of natural and synthetic types. Natural and synthetic polymers play a massive role in everyday life, and a life without polymers really does not exist.

Previously, it was believed that polymers could only be prepared through addition polymerization. The mechanism of the addition reaction was also unknown and hence there was no sound basis of proposing a structure for the polymers. Studies by researchers resulted in theorizing the condensation polymerization. This mechanism became well understood, and the structure of the resultant polyester could be specified with greater confidence.

In 1941/1942 the world witnessed the infancy of polyethylene terephthalate or better known as the polyester. A decade later for the first time polyester or cotton blends were introduced. In those days Terylene and Dacron (commercial names for polyester fibers) were miracle fibers but still overshadowed by nylon. Not many would have predicted that decades later, polyester would have become the world's inexpensive, general purpose fibers as well as becoming a premium fiber for special functions in engineering textiles, fashion, and many other technical end uses. From the time nylon and polyester were first used there have been an amazing technological advances that have made them so cheap to manufacture and widely available.

These developments have made the polymers such as polyesters contribute enormously in today's modern life. One of the most important applications is the furnishing sector (home, office, cars, aviation industry, etc.) which benefits hugely from the advances in technology. There are a number of requirements for a fabric to function in its chosen end use, for example, resistance to pilling and abrasion, as well as dimensional stability. Polyester is now an important part of upholstery fabrics. The shortcomings attributed to the fiber in its early days have mostly been overcome. Now it plays a significant part in improving the life-span of a fabric, as well as its dimensional stability, which is due to its heat-setting properties.

About half century has passed since synthetic leather, a composite material completely different from conventional ones, came to the market. Synthetic leather was originally developed for end-uses such as the upper of shoes. Gradually other uses such as clothing steadily increased the production of synthetic leather and suede. Synthetic leathers and suede have a continuous ultrafine porous structure comprising a three-dimensional entangled nonwoven fabric and an elastic material principally made of polyurethane. Polymeric materials consisting of the synthetic leathers are polyamide and polyethylene terephthalate for the fiber and polyurethanes with various soft segments, such as aliphatic polyesters, polyethers, and polycarbonates for the matrix.

New applications are being developed for polymers at a very fast rate all over the world at various research centers. Examples of these include electroactive polymers, nano products, robotics, etc. Electroactive polymers are special types of materials that can be used, for example, as artificial muscles and facial parts of robots or even in nano robots. These polymers change their shape when activated by electricity or even by chemicals. They are light weight but can bear a large force which is very useful when being utilized for artificial muscles. Electroactive polymers together with nanotubes can produce very strong actuators. Currently research works are carried out to combine various types of electroactive polymers with carbon nanotubes to make the optimal actuator. Carbon nanotubes are very strong, elastic, and they conduct electricity. When they are used as an actuator, in combination with an electroactive polymer, the contractions of the artificial muscle can be controlled by electricity. Already works are under way to use electroactive polymers in space. Various space agencies are investigating the possibility of using these polymers in space. This technology has a lot to offer for the future, and with the ever-increasing work on nanotechnology, electroactive materials will play a very important part in modern life.

This book presents some fascinating phenomena associated with the remarkable features of high performance polymers and also provides an update on applications of modern polymers. It offers new research on structure–property relationships, synthesis and purification, and potential applications of high performance polymers. The collection of topics in this book reflects the diversity of recent advances in modern polymers with a broad perspective that will be useful for scientists as well as for graduate students and engineers. The book helps to fill the gap between theory and practice. It explains the major concepts of new advances in high performance polymers and their applications in a friendly, easy-to-understand manner. Topics include high performance polymers and computer science integration in biochemical, green polymers, molecular nanotechnology and industrial chemistry. The book opens with a presentation of classical models, moving on to increasingly more complex quantum mechanical and dynamical theories. Coverage and examples are drawn from modern polymers.

— **Abbas Hamrang, PhD, and Bob A. Howell, PhD**

Section 1: Foundations of High Performance Polymers

CHAPTER 1

COMMERCIALLY PRODUCED AND MODIFIED STARCHES: PART I

A. I. SERGEEV, N. G. SHILKINA, L. A. WASSERMAN, and
H. STAROSZCZYK

CONTENTS

1.1 Chemicaly Modified Starches...2
    1.1.1 Introduction...2
    1.1.2 Experimental..3
    1.1.3 Discussion..4
    1.1.4 Conclusions..11
Keywords 11
References..11

1.1   CHEMICALY MODIFIED STARCHES

1.1.1   INTRODUCTION

In recent years, there has been increased interest in starch chemical modification reactions influenced by microwave irradiation [1–4]. The microwave irradiation using in the solid state etherification of starch with inorganic acids, there salts, anhydrides and oxides gave fast and uniform
heating, better selectivity, less byproducts, increasing the reaction rate
[1]. Such modification produced anionic starches as a negative charge on
them. The latter are essential for intermolecular interactions with other
macromolecules which lead, for instance, to the polysaccharide–protein
complexes formation. The resulting binary systems are known as biodegradable materials. Recently, natural biopolymers have been filled with
layered silicates to improve their functional properties. They might be
considered as environmental friendly components. The detailed synthesis
of starch silicates described in [3, 4]. The physical and structural properties of starch strongly depend on molecular interaction with water. The objective of this study was the investigation of the starch chemical modification effects on the state of water and water mobility during gelatinization.

Different techniques have been used for the study of water–starch interaction. Differencial scanning calorimetry (DSC) has been extensively
used for investigation of gelatinization, glass, and melting transition [5,
6]. X-ray diffraction method has been used for crystallization monitoring
[7], small angle neutron scattering (SANS)—of probe water inside starch
granules during gelatinization [8]. In wide spectrum of nuclear magnetic
resonance (NMR) methods, NMR-relaxation is undoubtedly the most suitable technique for studying the dynamic state of water in such heteroge-
neous systems as starch [9-13].

A lot of scientific methods which used in the recently works [3, 4] gave
a detailed functional properties of starch silicates. It is possible that the investigation of molecular dynamic of water interaction with such starches,
the quality, and quantity of unfrozen water will give an additional understanding of nature of these properties.

1.1.2 EXPERIMENTAL

The investigated starches were:

1. Native potato starch (PS) (Łomża, Poland)
2. Conventionally heated PS (2hrs at 100°C) (PSC)
3. Conventionally heated solid state mixture, PS + silicic acid (1:1 molar ratio) (PSCSi)
4. Microwave irradiated PS (30min 450W) (PSW)
5. Microwave irradiated the solid state mixture PS + H_2SiO_3 (1:1 molar ratio) (PSWSi)

The investigated samples were accomplished essentially as described by [4].

DSC

The differential scanning calorimeter DSC 204 F1 Phoenix (Netzsch, Germany) with Proteus Software on Windows was used to determine the frozen water (FW), the unfrozen water (UW), the additional unfrozen water (AUW), and the temperature of gelatinization (the peak melting) T_p in the starch–water systems. All samples were prepared at 1:1 ratio of starch to distilled water contents and stored for 24hrs in sealed aluminum pans. After experiment, the aluminum pans were pierced and dried out to the constant weight. DSC experiment consisted of several sets:

1. Sample cooling to –40°C
2. Scanning from –40 to +80°C at heating rate of 10°C/min (first scan)
3. Sample re-cooling to –40°C
4. Rescanning from –40 to +80°C at heating rate of 10°C/min (second scan)

The UW was calculated from the following equation:

$$UW = M_{com} - FW$$

where, M_{com}—water content arising from sample drying (g water/g dry starch),

FW—the frozen water, which is taken from the area under the ice melt-ing endothermically and heat of fusion of ice is 334.7J/g.

The AUW arising from gelatinization was calculated as difference be-tween FW of sequential scans:

$$AUW = FW_{1st\ scan} - FW_{2nd\ scan}$$

All measurements were done in triplicate.

NMR-RELAXATION METHOD

NMR-relaxation measurements were performed on Bruker Minispec PS 120 spectrometer operating at a proton resonance frequency of 20MHz.

The samples were placed in 5mm tube with a tefton plug inserted to prevent evaporation. The probe was equipped with a home made heat-ing control system (±0.5°C) and the measurements were taken at different temperatures ranging from 25 to 95°C. The sample had a thermal equili-bration for 10min at each temperature before measurement.

Spin-spin relaxation times (T_2) of water protons were measured using the Care-Purcell-Meiboom-Gill (CPMG) pulse sequence with 90°–180° pulse spacing (τ)—0.02ms, inter-pulse sequence delay—10s, number of points on curve—150. All data points of relaxation curve were acquired after each 10th echo as an average of 16 repetitions. Relaxation curves ob-tained from CPMG pulse sequence were analyzed by the MULTY T_2 pro-gram (Bruker), using a multiexponential model. All measurements were done in triplicate.

1.1.3 DISCUSSION

In a starch–water mixture, the water of hydration can be found near the surface of biopolymers. The water molecules can interact with macromol-ecules of polymers and could remain unfrozen at subzero temperature. The extent of hydration depends on several factors such as the distance of water molecules from surface of polymers, charge of the polymers, and density of hydrogen bonds. The structural function of UW is mainly driven by the state of polymer chains. If hydration water should migrate to form ice, rearrangement of the polymer structure would be necessary to adjust to the loss of those water molecules. But, this situation is difficult to realize because of the considerable energy required to change the polymer conformation at subzero temperature.

Therefore, sufficient amount of UW is retained in the system. In work [9], the authors showed that the total water content influenced the gel structure and altered the magnitude and extent starch–water interactions. They found that maximum AUW for a water content of several starches corresponded to a semi-solid saturated suspension (about 1:1 weight ratio) with a little or no additional supernatant water. In this condition, the extra-granule water has been proposed to exist as thin layer between closely packed granules [14]. The gel structure of those concentrated starch systems has also been described as tightly packed system of swollen granules with thin layer of amylose gel between the granules [15]. It is proposed that such a structure would be characterized by a short distance between hydration water and starch polymers and high local volume density of hydrogen bonds. As the initial water content increases, AUW of the systems tended to decrease the indicating reduction in starch–water interaction. All the starch types despite of the different amylose content, show relatively similar trend of the change in AUW as a function of initial water content [9]. This was pointed out of important role of amylopectin in interaction between water and gelatinized starches. Our results of DSC investigations of starches during gelatinization are present in Table 1.1.

TABLE 1.1   The amount of unfrozen water after gelatinization (UW), Additional unfrozen water after gelatinization (AUW) and temperature of gelatinization (Tp) for different starch–water mixtures.

	Unfrozen water after gelatinization g H_2O/g of dry starch (UW)	Additional unfrozen water after gelatinization g H_2O/g dry starch (AUW)	Gelatinization temperature (T_p peak) (°C)
PS	0.389 ± 0.009	0.021 ± 0.005	67.8 ± 0.2
PSC	0.382 ± 0.010	0.021 ± 0.004	67.5 ± 0.3
PSW	0.414 ± 0.012	0.039 ± 0.007	63.7 ± 0.8
PSCSi	0.360 ± 0.011	0.003 ± 0.002	64.3 ± 1.0
PSWSi	0.347 ± 0.012	0.013 ± 0.005	63.1 ± 0.7

AUW of native potato starch are close to those in the literature 0.38g H_2O/g of dry starch [16] and 0.43g H_2O/g of dry starch [9]. The amount of UW increases for all the samples after gelatinization. The convectional heating does not change AUW but microwave irradiation increases AUW almost twice. In work [4], it was shown that the changes in the granules structure induced by irradiation and etherification of starch weakened inter- and intra-molecular hydrogen bonds, facilitating the diffusion of water inside the granules. But on the other hand, the granules held together by cross linking became less water soluble. It has also provided granule agglomeration.

All these factors may be the reason of decrease in AUW for PSWS and PSCSi in our experiment. The microwave irradiation of starch without silicic acid led to deterioration of granule structure while its silication inhibited the granularity inspite of granule swelling. The decrease of gelatinization temperature T_p (Table 1.1) is also the result of crystal structure deterioration. Our results show that the change of AUW can reflect the destruction process in starch granules.

NMR RELAXATION

Proton transverse relaxation curves deconvolution for ungelatinized starch–water mixture samples reveal two distinct water population. The population of less mobile W_a and the more mobile W_b water have spin-spin relaxation times T_{2a} and T_{2b} respectively, only W_b is dependent on the initial water content (data not shown). It is possible to suggest that the less mobile water fraction W_a occupies the more constant environment, probably the interior of granule, and W_b—the exterior of granule.

The same model of starch–water interaction has been reported [14]. It was based on the assumption that the water within different compartments is distinguishable in an NMR relaxation study only if the rate of diffusive exchange of water molecules between two distinct regions is low as compared to their intrinsic relaxation rates. By investigating the temperature dependence of T_2 of water protons for saturated packed bed (1.1g water/g of dry starch) ungelatinized potato starch granules, three distinct water populations assigned to extra-granular water, water in the amorphous growth rings, and water in the amorphous regions of the crystalline lamellae have been identified. T_2 of the first population is about 50ms. The other two populations representing intra-granular water are in the fast diffusion exchange and have T_2 about 8ms.

As the system is cooled down to 4°C, the populations of intra-granular water enter slow, exchange regime and two T_2 appeared: the same water proton population for several starches discussed in work [9]. In [10] the authors reported about four populations for rice starch water mixture. But the population with the longest T_2 (1s) might be a result of water excess in samples (3:1) g H_2O/g of dry starch.

The identity of experimental conditions in these works gives us the ground to consider our assumptions of water population's location fair.

Dynamic redistribution of extra- and intra-granule water and the change of relaxation times T_2 in starch–water samples during gelatinization are demonstrated in Figure 1.1 and 1.2. The results can be divided into two groups:

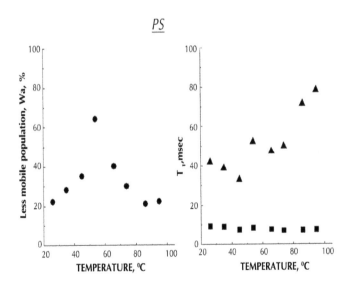

FIGURE 1.1 (*Continued*)

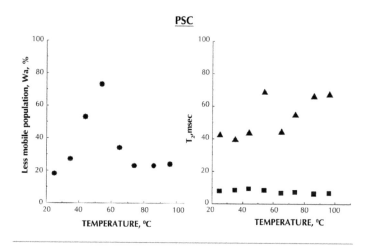

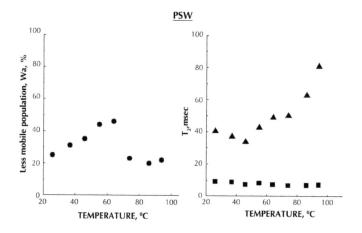

FIGURE 1.1   Changes in less mobile proton population Wa (•) and spin-spin relaxation times T_{2a} (■) and T_{2b} (▲) in starches during gelatinization.

1. PS, PSC, and PSW: The heating causes structural disruption of starch granules, water interaction with biopolymers occurs

easier and the W_a increases. After the partial gelatinization, some of the structural barriers are destroyed leading to a faster exchange of water molecules between different regions, as shown by reduction in less mobile water fraction W_a.

It was reported in the work [11] that such changes can reflect the increased mobility of starch polymer chains because the exchangeable water protons could be strongly influenced by the chemical exchange with hydroxyl protons in biopolymer. They pointed out that the increase of inter-granular water (W_a) with temperature reflected the swelling of granules. Above the melting transition temperature (62°C), W_a decreased as amylose diffused out of the granule. The difference of W_a in melting point for different samples (W_a maximum for PS, PSC, and PSW) can be explained by more destruction influence of microwave irradiation compared to convectional heating.

2. PSCS and PSWSi: W_a for the chemically modified starch at t = 25°C are approximately four times more than for native potato. It is necessary to note that it contradicts with UW decrease for the same samples. Possibly, it reflects the considerable reduction of extent of water interaction with biopolymers.

The morphological proton changes of granules of siliciated starch and DSC data which presented in [4] described the deterioration of granular structure. On the one hand the inter- and intra-molecular hydrogen bounds were weakened on the siliciation and granules could absorb water more easily, but on the other hand the granules held together by cross linking became less water soluble [4].

W_a of PSCSi and PSWSi monotonously decreases with heating. There are no considerable changes in gelatinization area. Presumably, it reflects the structural changes that chemically modified starches have.

Spin-spin relaxation times T_{2a} and T_{2b} have the same behavior during the heating for all the samples. T_{2a}—approximately constant, T_{2b}—increases with heating. This increase confirms our assumption of accessory, this population to inter-granular water in which dipole-dipole mechanism of relaxation prevails.

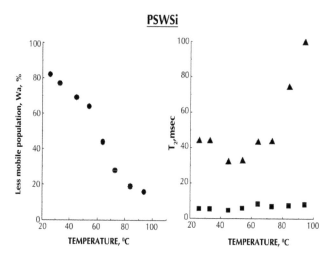

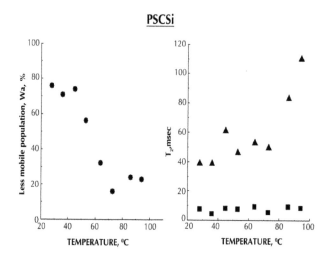

FIGURE 1.2   The changes in less mobile proton population Wa (●) and spin-spin relaxation times T_{2a} (■) and T_{2b} (▲) in chemically modified starches during gelatinization.

1.1.4   CONCLUSIONS

The hydration (AUW) of silicate starches during gelatinization decreased as compared to that of native potato starch. Amount of UW for ungelitinized silicate starches are less than in control samples. At the same time, the less mobile water population W_a in experiment is three times more than in control. Possibly, it reflects the reduction interaction energy of water with starch. These changes just as the gelatinization temperature decrease can be a consequence of starch destruction under microwave irradiation and chemical modification.

KEYWORDS

- **Differencial scanning calorimetry**
- **Frozen water**
- **Spin-spin relaxation**
- **Unfrozen water**
- **X-ray diffraction**

REFERENCES

1. Kappe, C. O. & Dallinger, D. (2009). Controlled microwave heating in modern organic synthesis: Highlights from 2004–2008 literature. *Molecular Diversity, 13,* 71–193.
2. Staroszczyk, H. (2009). Microwave-assisted boration of potato starch. *Polimery, 54,* 31–41.
3. Staroszczyk, H. (2009). Microwave-assisted silication of potato starch. *Carbohydrate Polymers, 77,* 506–515.
4. Staroszszyk, H & Janas, P. (2010). Microwave-assisted preparation of potato starch, silicated with silicic acid. *Carbohydrate Polymers, 81,* 599–606.
5. Donovan, J. W. (1979). Phase transition of the starch water system. *Biopolymers, 18,* 263–275.
6. Blanshard, J. M. V. (1987). Starch granule structure and function: A physicochemical approach. In T. Gallard (Ed.), *Starch: Properties and potential* (pp. 16–54). Chichester: Wiley.

7. Le Bail, H., Bizot, M., Ollivon, G., Koller, C., Bourgaux, & Buleon, A. (1999). Monitoring the crystallization of amylose–lipid complexes during maize starch melting by synchrotron X-ray diffraction. *Biopolymers, 50,* 99–110.
8. Jenkins, P. & Donald, A. M. (1998). Gelatinization of starch: A combined SAXS/WAXS/DSC and SANS study. *Carbohydrate Research, 308,* 133–147.
9. Tananuwong, K., Reid, D. S. (2004). DSC and NMR relaxation stages of starch–water interactions during gelatinization. *Carbohydrate Polymers, 58,* 345–358.
10. Ritita, M., Gianferri, R., Bucci, R., & Brosio, E. (2008). Proton NMR relaxation study of swelling and gelatinization process in rice starch–water samples. *Food Chemistry, 110*(1), 14–22.
11. Tang, H. R., Brun, A., & Hills, B. (2001). A proton NMR relaxation study of gelatinization and acid hydrolysis of native potato starch. *Carbohydrate Polymers, 46,* 7–18.
12. Choi, S. G. & Kerr, W. L. (2003). Water mobility and textural properties of native and hydroxypropilated wheat starch gels. *Carbohydrate Polymers, 51,* 1–8.
13. Choi, S. G. & Kerr, W. L. (2003). 1H NMR Studies of molecular mobility in wheat starch. *Food Research International, 36,* 34–348.
14. Tang, H. R., Godward, J., & Hills, B. (2000). The distribution of water in native starch granules—a multinuclear NMR study. *Carbohydrate Polymers, 43,* 375–378.
15. Garcia, V., Colonna, V., Bouchet, P., & Gallant, D. J. (1997). Structural changes of cassava starch granules after heating at intermediate water content. *Starch, 49,* 171–179.
16. Wootton, M. & Bamunuarachchi, A. (1978). Water binding capacity of commercially produced and modified starches. *Starch, 30,* 306–309.

CHAPTER 2

COMMERCIALLY PRODUCED AND MODIFIED STARCHES

S. Z. ROGOVINA, K. V. ALEKSANYAN, and E. V. PRUT

CONTENTS

2.1 Biodegradable Compositions of Starch 14
    2.1.1 Introduction.. 14
    2.1.2 Experimental 16
    2.1.3 Conclusion 31
Keywords 31
References... 32

2.1   BIODEGRADABLE COMPOSITIONS OF STARCH

2.1.1   INTRODUCTION

Because of constantly increasing production of polymers leading to environmental pollution, the problem of their utilization is becoming more and more urgent. This problem can be solved by creation of biodegradable polymer materials which can be decomposed under the environmental conditions to give harmless products for nature (H_2O, CO_2, biomass, and other natural compounds). Despite the fact that the cost of biodegradable polymers is higher than the cost of traditional synthetic polymers, they intensively lock in the consumer markets. The rise in application of biodegradable materials is primarily connected with irreplaceable consumption of the world reserves of oil products representing main raw material for production of synthetic polymers. With increase of production capacities of biodegradable polymer compositions and improvement of their production technology, their cost will gradually decrease and approach the cost of traditional synthetic polymers.

One of the most promising ways for producing partially or completely biodegradable polymer materials is based on the blending of synthetic and natural polymers. This concept of treating the enhancement of biodegradability of a polymer material via addition of a biodegradable component was proposed by Griffin in the early 1970s [1, 2]. Among such compositions, the blends based on polyolefins and renewable natural polysaccharides characterized by easy biodegradability are of special interest [3–6]. The methods of production of the blends based on polysaccharides with synthetic polymers are widely covered in the literature [7–9]. The properties of such compositions (thermal, sorption, and strength) are generally determined by both bond type between the components and their compatibility and character of forming morphology. Starch is a most popular polysaccharide, which is added to synthetic polymers, therefore, the blends based on it are the most extensively studied [10–12]. The production of packaging packets for stores was first launched in Europe in 1974 on the basis of starch–PE blends. Similar materials are also popular in production of biomedical items [13]. Starch has advantages such as low cost, wide availability, and total compostability without toxic residues, so the blends on its basis can be used as substitutes to petroleum-based plastics [14].

Starch is widely used as a filler for production of reinforced plastics [15, 16]. Even small amount of starch (6–30%) mixed with synthetic polymer increases the biodegradability of the final products [17]. However, because of differences in chemical structure and hydrophilic characteristics of these two components, the polymer blends have low compatibility that leads to deterioration of mechanical properties. This problem can be solved by chemical modification of one of the components with the aim to introduce functional groups capable to react with second component [18]. Thermoplastic blends of starch with synthetic polymers are obtained with the use of starch plasticized by water, glycerol, and so on [19]. In this case the components are usually mixed in an extruder at ~150°C providing well gelatinization of starch [20].

Since there are, as a rule, no common solvents for natural polysaccharides and synthetic polymers, the production of their blends by solid-phase mixing is of great interest. Besides, due to the absence of solvents, the solid-phase mixing has an advantage as high ecological safe method.

Among various methods of polymer mixing in solid phase (e.g., centrifugal and ball mills), the extrusion mixing is quite promising one. This method, where the principle of joint action of temperature, pressure, and shear deformations on material is realized, is widely studied by the Russian scientist [21]. This method is based on destruction of the initial components under action of uniform compression and shear deformations. The physical principle underlying the method proposed is in the fact that energy stored in material under action of pressure and shear deformations is realized in formation of new surface, small shearing force is sufficient. As a result of such action, the powders of different dispersion are formed. It should be noted that the grinding process is not a gradual breakage of the material to smaller particles, but sharp process in a restricted space that points to spontaneous destruction and allows one to assume the branched mechanism of breakdown site development. Use of this method allows one to obtain uniform homogeneous powder blends in the absence of common solvents.

The aim of this work is to produce binary biodegradable blends of starch with low density polyethylene (LDPE) (T_m = 108°C, M_n = 23,000,

MFI = 2g/10min (190°C, 2.16kg)), their ternary blends with poly(ethylene oxide) (PEO) of various molecular weight (low-molecular PEO_{lm}, $M = 3.5 \times 10^4$ and high-molecular PEO_{hm}, $M = 5 \div 6 \times 10^6$) and natural polysaccharide chitin (content of basic compound 87.2%, degree of deacetylation 0.045) and its deacetylated derivative chitosan (degree of deacetylation 0.87 and $M = 4.4 \times 10^5$) that significantly expands the possible fields of application of such compositions. Structure, biodegradability, and physico–mechanical properties of above listed blends were studied in detail.

2.1.2   EXPERIMENTAL

Starch was blended with various polymers at 150°C via single pass of components through a rotor disperser, in which material was subjected to joint action of temperature, pressure, and shear deformations. To create intensive shear deformations, the rotor disperser was equipped with a grinding head designed as a cam element rotating inside a channeled cylinder and a cooling jacket. Whatever the blend composition was, homogeneous fine-dispersed powders were formed after passing of polymers through the rotor disperser.

The systems of the following composition were obtained: starch–LDPE binary compositions with starch content from 20 to 50 wt.%; starch–LDPE–PEO ternary compositions containing 20 to 40 wt.% of starch, and starch–LDPE–chitin or chitosan ternary systems with 30 wt.% of starch. The obtained blends were fractionated with the use of a sieve set. Figure 2.1 shows the comparative histograms of binary and ternary compositions containing 30 wt.% of starch. As is seen from the given data, for starch–LDPE binary composition, the basic fraction (about 60%) is composed of particles 0.315–0.63mm in size. It should be noted that for binary composition, the content of fractions with a large particle size increases with the rise of starch content in blend (Table 2.1). Essentially, the same fraction distribution is observed for starch–LDPE–PEO_{lm} ternary composition. In this case, two fractions with the particle size from 0.09 to 0.63mm are basic, the fractions with particle size from 0.8mm appear. Somewhat different distribution was obtained for compositions based on two polysaccharides (e.g., starch and chitin). In this case, the basic fraction (above 70%) is composed of particles 0.09–0.315mm in size. At the same time, new fraction containing small particles 0.071–0.09mm in size appears. Thus, although the total content of polysaccharides increases from 50 to 60%, the powder with particles of finer size is formed during passing through the rotor that is likely to be connected with co-grinding process in this case [21].

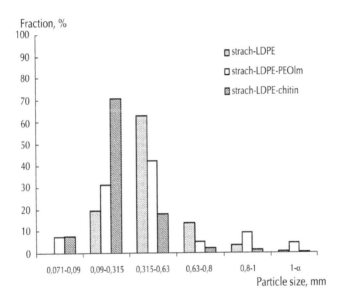

FIGURE 2.1 Histograms of particle size distribution of starch–LDPE (30:70 wt.%), starch–LDPE–PEO$_{lm}$ (30:50:20 wt.%), and starch–LDPE–chitin (30:40:30 wt.%) blends.

TABLE 2.1 Fractional composition of starch–LDPE blend at different component ratios.

Blend com-position	Compo-nent ratio (%)	Fraction with d (mm), (%)				
		0.071–0.09	0.09–0.315	0.315–0.63	0.63–0.8	0.8–1.0
	20:80	17.7	53.5	23.6	2.8	2.4
Starch–LDPE	30:70	0	19.6	62.8	13.8	3.8
	50:50	0	10.8	78.6	6.8	3.8

The fractional composition of starch–LDPE blend was determined by iodine reaction. For this purpose, starch was washed with hot water until the disappearance of blue color of water in the presence of iodine. The residual LDPE was dried, and the component ratios in fractions were calculated (Table 2.2). As is seen from the given data, the fractional composition is nearly consistent with the initial component ratio in the blend that testifies the homogeneous distribution of the blend components.

TABLE 2.2   LDPE content in fractions of starch–LDPE blend of different compositions determined by chemical method.

Fraction (mm)	Initial content of LDPE (wt.%)		
	80	**70**	**50**
0.071–0.09	88	-	-
0.09–0.315	89	81	76
0.315–0.63	87	87	76
0.63–0.80	-	88	65

Structure of the binary compositions was investigated by the X-ray diffraction analysis. The powder samples were studied on a Rigaku RU-200 Rotaflex diffractometer (Japan) with a rotating copper anode in the transmission mode (40kV, 140mA, CuK_α radiation with $\lambda = 0.1542$nm). Figure 2.2 shows the X-ray patterns of starch, LDPE, and their blend obtained under shear deformations. The X-ray pattern of starch exhibits a set of reflections at $2\theta = 5.7°$ and $16.9°$ characteristic for crystalline structure. In the X-ray pattern of starch–LDPE blend subjected to shear deformations, these reflections and characteristic peak of LDPE are weaker; that is, the amorphization of the initial polymers takes place. Thus, the mixing of polymers under shear deformations lead to decrease of their crystallinity degree.

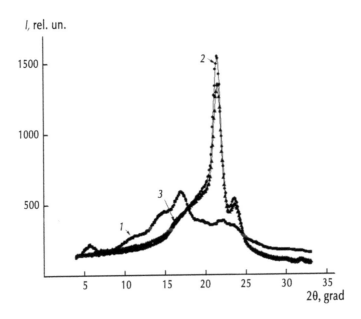

FIGURE 2.2   X-ray patterns of starch (*1*), LDPE (*2*), and starch–LDPE blend (30:70wt.%) obtained under action of shear deformations (*3*).

These results were confirmed by DSC data obtained during investigation of starch–LDPE compositions produced by mechanical mixing or under the action of shear deformations on a DSM-10m calorimeter. The weight of samples was in the range from 2 to 6mg, the heating rate was 16K/min. The melting enthalpies of individual LDPE and PEO, and binary and ternary blends were calculated per unit mass of LDPE and PEO, respectively, in blend.

It turned out that the position of peaks for LDPE on all three curves coincides, and peak areas differ between each other that is caused by both starch introduction and amorphization of blend obtained under shear deformations (Figure 2.3). As is seen from figure, the curve of starch–LDPE blend displays a single peak characteristic for LDPE. The position of this peak on the DSC curves for the blends corresponds to T_m of LDPE independently of the conditions of blend preparation. However, values of the melting enthalpies ΔH of blends prepared via mechanical mixing or under shear deformations differ from the values of LDPE ΔH. In this case, the value of melting enthalpy ΔH for blend prepared in the rotor disperser is lower than ΔH for blend obtained by mechanical mixing (Table 2.3). So, for starch–LDPE (30:70 wt.%) blend prepared under action of shear deformations, the melting enthalpy equals 43J/g, while for the blend obtained by mechanical mixing $\Delta H = 53$J/g, that is, the crystallinity degree is lower for the blend obtained in the rotor disperser. Consequently, one may state that LDPE amorphization occurs during the production of blends under the action of shear deformations.

TABLE 2.3 Basic parameters of melting (melting point T_m, temperature range of melting ΔT_m, and melting enthalpy ΔH) of LDPE and its blends with starch obtained under shear deformations (1) and by mechanical mixing (2).

System	T_m (°C)	ΔT_m (°C)	ΔH (mJ)	ΔH (J/g)
LDPE	109	92–126 34	155	52
Starch–LDPE (30:70wt.%) (1)	108.5	93–124 31	107	43
Starch–LDPE (30:70wt.%) (2)	108	93–125 32	128	53

Figure 2.3 shows the melting curves of LDPE, starch–LDPE (30:70 wt.%) binary composition, and starch–LDPE–PEO$_{lm}$ (30:50:20 wt.%) ternary composition. As is seen from the presented data, the melting curve of the binary blend displays a single peak (characteristic for LDPE) and the melting curve of the ternary blend displays two peaks (characteristic for LDPE and PEO). Change in the melting enthalpies of LDPE and PEO, in addition to the enthalpy change due to amorphization, which is constant for these blends, depends also on the blend composition. Based on this fact, the content of the blend components can be calculated. As is seen from Table 2.4, presenting the data for the ternary blends based on PEO$_{lm}$, the calculated values are close to the initial ratio of components. Thus, DSC can be used as an indirect method to calculate starch content in the ternary blends with two synthetic polymers.

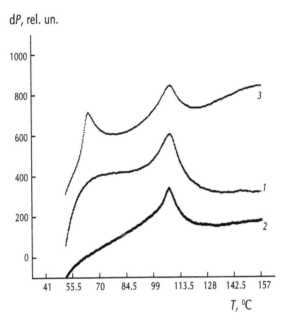

FIGURE 2.3 Melting curves of LDPE (*1*), starch–LDPE (30:70 wt.%) (*2*), and starch–LDPE–PEO$_{lm}$ (30:50:20 wt%) (*3*) compositions obtained by DSC.

TABLE 2.4 Composition of starch–LDPE–PEO$_{lm}$ blend determined by DSC.

System	Compo-nent ratio (wt.%)	ΔH (PEO$_{lm}$) (J/g)	ΔH (LDPE) (J/g)	Determined content of components (%)		
				PEO$_{lm}$	LDPE	Starch
Starch–LDPE–PEO$_{lm}$	20:60:20	23	31	17.8	55.8	26.4
	30:50:20	29	29	22.3	51.5	26.2
	40:40:20	31	28	23.5	49.0	27.5

The samples of initial powder starch, starch–LDPE (30:70 wt.%) blend, and cross-section of the film from starch–LDPE blend obtained after freezing in a liquid nitrogen were investigated with the use of scanning electron microscopy (SEM). The study was performed on SEM JEOL JSM-7001F at various magnifications. Electron micrographs show that the starch particles have ideal shape, and their size varies from 12 to 60µm (Figure 4a). From consideration of electron micrographs of the powder

starch–LDPE blend, it follows that starch particles are uniformly distributed in LDPE matrix (Figure 4b). Figure 4c–e illustrate LDPE bands of
starch particles in the cross-section of the film from starch–LDPE blend
that testifies some interfacial interaction.

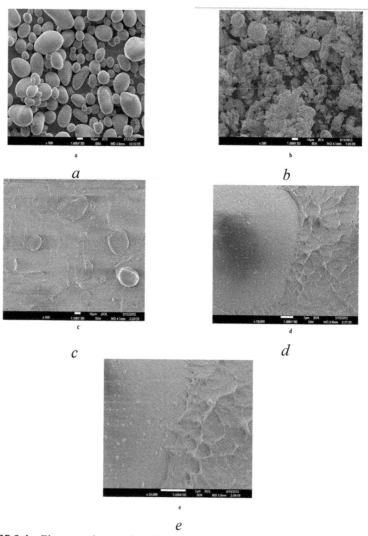

FIGURE 2.4 Electron micrographs of the initial starch (a), powder starch–LDPE (30:70
wt.%) blend (b) and cross-section of the film obtained from starch–LDPE (30:70 wt.%)
blend (c–e) at various magnifications: a, b, c—X500; d—X10,000; e—X20,000.

Mechanical tests were performed on films obtained from binary and ternary powder compositions by pressing. Figure 2.5 shows the comparative dependences of elastic modulus E, ultimate tensile strength σ_b, and elongation at break ε_b of starch–LDPE binary composition and ternary compositions of starch with LDPE containing low-molecular and high-molecular PEO on starch content. It is seen from this figure that as starch content increases, the value of elastic modulus E rises for all systems. Maximal values were obtained for the compositions containing high-molecular PEO. The character of curves of E dependence on starch content of the binary and ternary compositions with low-molecular PEO containing up to 30 wt.% of starch remains almost unchanged. With further increase of starch content, the value of elastic modulus for ternary blends sharply rises.

The value of ultimate tensile strength σ_b of starch–LDPE and starch–LDPE–PEO$_{hm}$ compositions changes insignificantly with increase of starch content, but it is obvious that the highest values are observed for the binary compositions. Ultimate tensile strength of starch–LDPE–PEO$_{hm}$ composition drops monotonically with increase of starch content, while value of elongation at break ε_b decreases for all compositions. If for starch—LDPE (curve 1) and starch–LDPE–PEO$_{hm}$ (curve 3) the drop of ε_b is not so substantial (on 20 and 14%, respectively), so for starch–LDPE–PEO$_{lm}$ (curve 2) ternary composition, the value of elongation at break sharply decreases with increase of starch content (on 60%). Absolute values of ultimate tensile strength σ_b and elongation at break ε_b of binary compositions are higher than for the ternary compositions based on two synthetic polymers. Such character is apparently connected with different morphology of the compositions obtained.

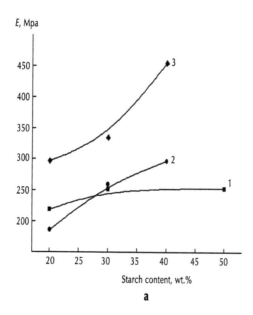

a

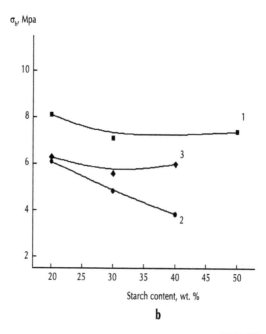

b

FIGURE 2.5   (*Continued*)

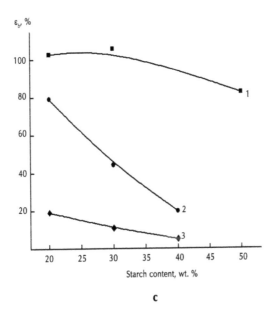

FIGURE 2.5 Dependence of elastic modulus E (a), ultimate tensile strength σ_b (b), and elongation at break ε_b (c) on starch content for starch–LDPE (*1*), starch–LDPE–PEO$_{lm}$ (*2*), and starch–LDPE–PEO$_{hm}$ (*3*) compositions.

Table 2.5 shows the dependences of mechanical parameters for binary and ternary compositions based on two synthetic polymers and two polysaccharides with the same starch content. From data presented, it follows that elastic modulus E of starch–LDPE–chitin or chitosan composition is significantly higher than for the rest systems, that is caused by increased rigidity of chitin and chitosan in comparison with other polymers. Similar dependence is observed for the ultimate tensile strength. As one would expect, the values of elongation at break ε_b of the ternary systems based on two polysaccharides is significantly lower than of ternary compositions based on two synthetic polymers and starch–LDPE binary compositions.

Thus, a change in the mechanical characteristics of binary blends based on starch with LDPE and ternary blends with introduction of PEO or chitin or chitosan depends on both blend composition and component nature.

TABLE 2.5 Dependence of elastic modulus E, ultimate tensile strength σ_b, and elongation at break ε_b on blend composition.

Composition (wt.%)	E (MPa)	σ_b (MPa)	ε_b (%)
Starch–LDPE (30:70)	290 ± 10	7.1 ± 0.2	105 ± 5
Starch–LDPE–PEO$_{lm}$ (30:50:20)	260 ± 10	4.85 ± 0.1	4.44 ± 0.3
Starch–LDPE–chitin (30:40:30)	968 ± 54	10.5 ± 20	2.1 ± 0.2
Starch–LDPE–chitosan (30:40:30)	796 ± 8	11.5 ± 0.3	3.05 ± 0.1

Thermal stability is one of the main characteristics of polymer compositions determining possible areas of their application. Therefore, investigation of the dependence of blend thermal stability on its composition is of both scientific and practical interest. Thermogravimetric analysis of the individual polymers and their blends was carried out on a synchronic thermal analyzer STA 449 F3 Jupiter (NETSCH, Germany) over the temperature range of 30–560°C with the rate of temperature change of 10K/min in an argon atmosphere (rate of gas consumption was 10mL/min). Figure 2.6 shows TGA curves of individual polymers and their binary and ternary blends. As is seen from Figure 2.6a, introduction of LDPE leads to increase of the composition thermal stability. Apart from thermal stability, photooxidation is also an important factor determining polymer system stability. It is known that photodegradation of such synthetic polymers as polyethylene and polypropylene leading to formation of low-molecular fractions causes increase of polymer biodegradability [22, 23]. Comparison of TGA curves of starch–LDPE binary composition before and after photooxidation (ISO 4892-1) showed that the temperature of the degradation onset for the irradiated blend is lower than for the blend before photooxidation.

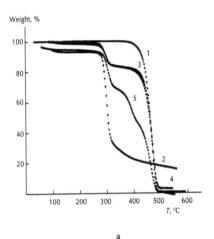

a

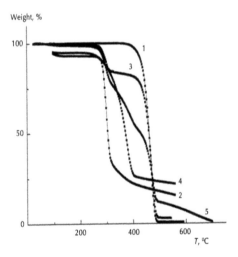

FIGURE 2.6 TGA curves of individual polymers and their blends: LDPE (*1*), starch (*2*), starch–LDPE (50:50 wt.%) before irradiation (*3*); (a) starch–LDPE (50:50 wt.%) after irradiation (*4*), starch–LDPE–PEO$_{lm}$ (40:40:20 wt.%) (*5*); (b) chitin (*4*), starch–LDPE–chitin (30:40:30 wt.%) (*5*).

Figure 6a also shows TGA curve of starch–LDPE–PEO$_{lm}$ (30:50:20 wt.%) ternary composition. It is visible that addition of PEO leads to significant decrease of the system thermal stability compared to the binary systems. Figure 6b illustrates TGA curves of LDPE, starch, chitin, and TGA curve of the ternary blend based on two polysaccharides, for example, starch–LDPE–chitin (30:50:30 wt.%). The comparison of TGA curves of binary and ternary compositions based on starch showed that the degradation temperature onset is approximately the same. In the temperature range from 300 to 420°C, the binary composition reduces in weight on 7–10%, while the ternary composition based on two polysaccharides reduces in weight by 30–34%.

Data of thermogravimetric analysis allows one to state that thermal destruction of the compositions based on two synthetic polymers and two polysaccharides, where their total fraction is up to 60%, proceeds more intensively than degradation of the binary compositions based on starch with LDPE, where this factor does not exceed 50%.

The estimation methods of polymer composition biodegradability mainly based on analysis of waste products of the microorganisms are described in detail and introduced into standards of various countries [24–27]. They involves artificial infection of the samples by fungus test cultures (e.g., *Aspegilus*, *Penicillium*, etc.), which are more active in relation to a lot of classes of polymers [28].

Among various methods on investigation of the polymer composition biodegradability, generally accepted tests on fungus resistance and tests on biodegradation under natural conditions, according to which the samples placed into wet soil were used.

The biodegradability of starch–LDPE binary blends was determined by the intensity of mold fungus growth on the samples. Various fungi from the All-Russia Collection of Microorganisms were used (Table 2.6). According to the tests on fungus resistance of the binary compositions based on starch, starch—LDPE (50:50 wt.%) blend shows maximum intensity of mold fungus growth, measured at 5 (Figure 2.7). The growth of fungi covering more than 90% of the surface is clearly seen by the naked eye. The films prepared from these blends have no fungus resistance and contain nutrients favoring fungi growth in the presence of mineral pollutants.

TABLE 2.6 Types and concentration of spores from All-Russia Collection of Microorganisms.

Types of spores	Amount of spores	Concentration of spores (N/sm³)
Aspergillus niger van Tieghem	1,550	7.8×10^6
A. terreus Thom	2,050	1.0×10^7
A. oryzae (Ahlburg) Cohn	1,200	6.0×10^6
Chaetomium globosum Kunze	200	1.0×10^6
Paecilliumces variotii Bainier	1,650	8.3×10^6
Penicillium funiculosum	435	2.1×10^6
P. chrysogenum Thom	715	3.6×10^6
P. cyclopium Westling	465	2.3×10^6
Trichoderma viride Persoon: Fries	800	4.0×10^6

 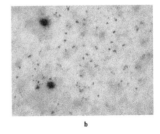

a   b

FIGURE 2.7 Micrographs of surface of starch–LDPE (50:50 wt.%) composition: **a**—original film, **b**—film infected with fungi spores for 28 days.

However, it turned out that this composition is not completely biodegradable. During holding the films in soil for several months, they reduced insignificantly in weight but in this case, the films became more fragile with dark spots corresponding to the biodegradation areas.

The curves of weight loss for starch–LDPE–chitin or chitosan ternary compositions placed in soil are given in Figure 2.8. As illustrated in this figure, both blends reduce in weight up to 25%. It should be noted that the weight loss rate for all systems explored was the same during first 50 days, but on further holding in soil the rate of biodegradation for starch–LDPE–chitin composition was higher than for the similar system containing chitosan.

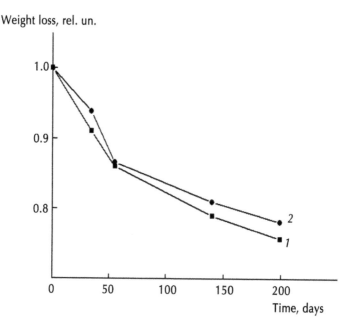

FIGURE 2.8   Weight loss curves of films from starch–LDPE–chitin (*1*) and starch–LDPE–chitosan (*2*) (30:40:30 wt.%) compositions.

Table 2.7 shows data characterizing the difference of the mechanical characteristics of ternary compositions based on starch before and after holding in soil. The data presented testify that the samples reduce not only in weight during the investigation, but also in strength.

TABLE2.7 Change of mechanical characteristics of starch–LDPE–chitin or chitosan compositions after holding in soil.

Composition (30:40 30 wt.%)	ΔE (MPa)	$\Delta\sigma_b$ (MPa)	$\Delta\varepsilon_b$ (%)
Starch–LDPE–chitin	572 ± 32.5	5.5 ± 1.6	0.05 ± 0.2
Starch–LDPE–chitosan	571 ± 25.5	8.8 ± 0.3	1.05 ± 0.1

2.1.3   CONCLUSION

Biodegradable compositions based on starch with synthetic polymers and some other polysaccharides were obtained in the rotor disperser under conditions of shear deformations that allows one to obtain highly homogeneous powders. The results of X-ray analysis and DSC showed that the mixing of polymers under shear deformations lead to decrease of their crystallinity degree. Thermogravimetric analysis showed that the introduction of LDPE leads to increase of the binary system thermal stability, while the photooxidation and especially addition of the third component leads to significant decrease of the system thermal stability. The mechanical properties and biodegradability of these blends generally depend on the composition of blend. Both addition of PEO and other polysaccharide (chitin or chitosan) to starch–LDPE compositions leads to an increase of the material biodegradability with insignificant deterioration of the mechanical parameters. Thus, the solid-phase ecologically safe method of polymer mixing permits one to produce the biodegradable blends based on starch and natural or synthetic polymers with quite high mechanical characteristics that allows one to use them in considerably important fields as the production of packaging materials, domestic articles, medicine, and so on.

KEYWORDS

- Biodegradable polymer
- Hydrophilic
- Low density polyethylene
- Starch
- Thermogravimetric analysis

REFERENCES

1. Griffin, G. J. L. (1974). Biodegradable fillers in thermoplastics. *Advances in Chemistry Series, 134,* 159–170.
2. Griffin, G. J. L. (1973). *British Pathological Application, 55195,* 73.
3. Vasnev, V. A. (1997). Biodegradable polymers. *Polymer Science Series B, 39,* 474–485.
4. Arvanitoyannis, I. (1999). Totally and partially biodegradable polymer blends based on natural and synthetic macromolecules: Preparation, physical properties, and potential as food packaging materials. *Journal of Macromolecular Science—Reviews in Macromolecular Chemistry and Physics, 39*(2), 205–271.
5. Albertson, A. C. & Karlsson, S. (1994). Chemistry and biochemistry of polymer biodegradation. In G. J. L. Griffin (Ed.), *Chemistry and technology of biodegradable polymers* (pp. 7–17). Glasgow: Blackie Academic and Professional.
6. Bastioli, C. (2005). *Handbook of biodegradable polymers.* Shawbury: Rapra Technology ltd.
7. Huneault, M. A. & Li, H. (2012). Preparation and properties of extruded thermoplastic starch/polymer blends. *Journal of Applied Polymer Science, 126*(S1), E96–E108.
8. Xiong, H. G., Tang, S. W., Tang, H. L., & Zou, P. (2008). The structure and properties of a starch-based biodegradable film. *Carbohydrate Polymers, 71*(2), 263–268.
9. Kaseem, M., Hamad, K., & Deri, F. (2012). Preparation and studying properties of polybutene-1/thermoplastic starch blends. *Journal of Applied Polymer Science, 124*(4), 3092–3098.
10. Aburto, J., Thiebaud, S., Alric, I., Borredon, E., Bikiaris, D., Prinos, J., & Panayiotou, C. (1996). Properties of octanoated starch and its blends with polyethylene. *Carbohydrate Polymers, 34*(1), 101–112.
11. Thakore, I. M., Desai, S., Sarawade, D. D., & Davi, S. (2001). Studies on biodegradability, morphology and thermo-mechanical properties LDPE/modified starch blends. *European Polymer Journal, 37*(1), 151–160.
12. Jang, B. C., Huh, S. Y., Jang, J. G., & Bae, Y. C. (2001). Mechanical properties and morphology of the modified HDPE/starch reactive blend. *Journal of Applied Polymer Science, 82*(13), 3313–3320.

13. Sionkowska, A. (2011). Current research on the blends of natural and synthetic polymers as new biomaterials: Review. *Progress Polymer Science, 36,* 1254–1276.
14. Xie F., Halley P. J., & Avérous, L. (2012). Rheology to understand and optimize processibility, structures and properties of starch polymeric materials. *Progress Polymer Science, 37,* 595–623.
15. Bagheri, R. (1999). Effect of processing on the melt degradation of starch-filled polypropylene. *Polymer International, 48*(12), 1257–1263.
16. Evangelista, R. L., Nikolov, Z. L., Sung, W., Jane, J., & Gelina, R. J. (1991). Effect of compounding and starch modification on properties of starch-filled low density polyethylene. *Industrial and Engineering. Chemistry. Research, 30*(8), 1841–1846.
17. Röper, H. & Koch, H. (1990). The role of starch in biodegradable thermoplastic materials. *Starch–Stärke, 42,* 123–130.
18. Pimpan, V., Ratanajiajaroen, P., & Laithumtaweekul, L. (2007). Effect of chemical modification on the properties of cassava starch and medium density polyethylene blends. Abstracts of 9th European Symposium "Polymer Blends" . Palermo, September 9–12, p. 27.
19. Russo, M. A. L., O'Sullivan, C., Rounsefell, B., Halley, P. J., Truss, R., & Clarke, W. P. (2009). The anaerobic degradability of thermoplastic starch: Polyvinyl alcohol blends: Potential biodegradable food packaging materials. *Bioresource Technology, 100,* 1705–1710.
10. St.-Pierre, N., Favis, B. D., Ramsay, J. A., & Verhoogt, H. (1997). Processing and characterization of thermoplastic starch/polyethylene blends. *Polymer, 38*(3), 647–655.
11. Prut, E. V. (1994). Creep instability and fragmentation of polymers (Review). *Polymer Science Serial A, 36*(4), 601–607.
12. Stranger-Johannessen, M. (1979). Susceptibility of photodegradeed polyethylene to microbiological attack. *Journal of Applied Polymer Science, 35,* 415–421.
13. Cornell, J. H., Kaplan, A. M., & Roger, M. R. (1984). Biodegradability of photooxidized polyalkylenes. *Journal of Applied Polymer Science, 29*(8), 2581–2597.
14. Calmon-Decriaud, A., Bellon-Maurel, V., & Silvestre, F. (1998). Standard methods for testing the aerobic biodegradation of polymeric materials. Review and Perspectives. *Advance Polymer Science, 135,* 207–226.
15. Vikman, M., Itävaara, M., & Poutanen, K. (1995) Measurement of the biodegradation of starch-based materials by enzymatic methods and composting. *Journal of Polymers and the Environment, 3*(1), 23–29.
16. Urstadt, S., Augusta, J., Müller, R. -J., & Deckwer, W. -D. (1995). Calculation of carbon balances for evaluation of the biodegradability of polymers. *Journal of Polymers and the Environment, 3*(3), 121–131.
17. Kleeberg, I., Hetz, C., Kroppenstedt, R. M., Müller, R. -J., & Deckwer, W. -D. (1998). Biodegradation of aliphatic-aromatic copolyesters by *Thermomonospora fusca* and other thermophilic compost isolates. *Applied and Environmental Microbiology, 64,* 1731–1735.
18. Ryabov, S. V., Boiko, V. V., Kobrina, L. V. et al. (2006). Study and characterization of polyurethane composites filled with polysaccharides. *Polymer Science Serial A, 48*(8), 841–853.

CHAPTER 3

PRODUCTION AND APPLICATION OF POLYMER NANOCOMPOSITES: PART I

G. V. KOZLOV, K. S. DIBIROVA, G. M. MAGOMEDOV, BOB A. HOWELL, and G. E. ZAIKOV

CONTENTS

3.1 The Reinforcement Mechanism of Nanocomposites Polyethylene or Organoclay on Supra-Segmental Level 36

    3.1.1 Introduction... 36

    3.1.2 Experimental... 37

    3.1.3 Results and Discussion 37

    3.1.4 Conclusions... 41

Keywords 42

References... 42

3.1   THE REINFORCEMENT MECHANISM OF
NANOCOMPOSITES POLYETHYLENE OR ORGANOCLAY ON
SUPRA-SEGMENTAL LEVEL

3.1.1   INTRODUCTION

Very often, a filler (nanofiller) is introduced in polymers with the purpose of latter stiffness increase. This effect is called polymer composites (nanocomposites) reinforcement and it is characterized by reinforcement degree E_c/E_m (E_n/E_m), where E_c, E_n, and E_m are elasticity moduli of composite, nanocomposite, and matrix polymer, accordingly. The indicated effect significance results in a large number of quantitative models development, describing reinforcement degree: micromechanical [1], percolation [2], and fractal [3] ones. The principal distinction of the indicated models is the circumstance, that the first ones take into consideration the filler (nanofiller) elasticity modulus and the last two—donot. The percolation [2] and fractal [3] models of reinforcement assume that the filler (nanofiller) role comes in the modification and fixation of matrix polymer structure. Such approach is obvious enough, if to account for the difference of elasticity modulus of filler (nanofiller) and matrix polymer. So in the present paper, nanocomposites low density polyethylene/Na$^+$-montmorillonite, the matrix polymer elasticity modulus makes up 0.2GPa [4] and nanofiller—400–420GPa [5], that is, the difference makes up more than three orders. It is obvious that at such conditions organoclay strain is practically equal to zero and nanocomposites' behavior in mechanical tests is defined by polymer matrix behavior.

Lately it was offered to consider polymer's amorphous state structure as a natural nanocomposite [6]. Within the frameworks of cluster model of polymers amorphous state structure, it is supposed that the indicated structure consists of local order domains (clusters), immersed in loosely-packed matrix, in which the entire polymer free volume is concentrated [7, 8]. In its turn, clusters consist of several collinear densely-packed statistical segments of different macromolecules, that is, they are an amorphous analog of crystallites with stretched chains. It has been shown [9] that clusters are nanoworld objects (true nanoparticles–nanoclusters) and in case of polymers representation as natural nanocomposites, they play nanofiller role and loosely-packed matrix-nanocomposites—matrix role. It is significant that the nanoclusters dimensional effect is identical to the indicated effect for particulate filler in polymer nanocomposites—size decrease of both nanoclusters [10] and disperse particles [11] results in sharp enhancement of nanocomposite reinforcement degree (elasticity modulus). In connection with the indicated observations, the question arises: how organoclay introduction in polymer matrix influences the nanoclusters size and how the variation of the latter influences nanocomposite elasticity modulus value. The purpose of the present paper is to find the solution of these two problems using the example of nanocomposite linear low density polyethylene or Na^+-montmorillonite [4].

3.1.2 EXPERIMENTAL

Linear low density polyethylene (LLDPE) of mark Dowlex-2032 having melt flow index 2.0g/10min and density 926kg/m^3, that corresponds to crystallinity degree of 0.49, used as a matrix polymer. Modified Na^+-montmorillonite (MMT), obtained by cation exchange reaction between MMT and quaternary ammonium ions, was used as nanofiller MMT contents which make up 1–7mass% [4].

Nanocomposites linear low density polyethylene or Na^+-montmorillonite (LLDPE/MMT) were prepared by components blending in melt using Haake twin-screw extruder at temperature 473K [4].

Tensile specimens were prepared by injection molding on Arburg Allounder 305-210-700 molding machine at temperature 463K and pressure 35MPa. Tensile tests were performed by using tester Instron of the model 1137 with direct digital data acquisition at temperature 293K and strain rate ~$3.35 \times 10^{-3}s^{-1}$. The average error of elasticity modulus determination makes up 7%, yield stress—2% [4].

3.1.3 RESULTS AND DISCUSSION

For the solution of first from the indicated problems, the statistical segments number in one nanocluster n_{cl} and its variation at nanofiller contents change should be estimated. The parameter n_{cl} calculation consistency includes the following stages. At first, the nanocomposite structure fractal dimension d_f is calculated according to the equation [12]:

$$d_f = (d-1)(1+v) \qquad (3.1),$$

where, d is dimension of Euclidean space in which a fractal is considered (it is obvious that in our case $d = 3$), v is Poisson's ratio, which is estimated according to mechanical tests results with the aid of the relationship [13]:

$$\frac{\sigma_Y}{E_n} = \frac{1-2v}{6(1+v)} \qquad (3.2),$$

where, σ_Y and E_n are yield stress and elasticity modulus of nanocomposite, accordingly.

Then nanocluster's relative fraction φ_{cl} can be calculated by using the following equation [8]:

$$d_f = 3 - 6\left(\frac{\phi_{cl}}{C_\infty S}\right)^{1/2} \qquad (3.3),$$

where, C_∞ is the characteristic ratio, which is a polymer chain statistical flexibility indicator [14], S is macromolecule cross-sectional area.

The value C_∞ is a function of d_f according to the relationship [8]:

$$\tilde{N}_\infty = \frac{2d_f}{d(d-1)(d-d_f)} + \frac{4}{3} \qquad (3.4).$$

The value S for low density polyethylenes is accepted equal to 14.9Å^2 [15]. Macromolecular entanglements cluster network density v_{cl} can be estimated as follows [8]:

$$v_{cl} = \frac{\phi_{cl}}{C_\infty l_0 S} \qquad (3.5),$$

where, l_0 is the main chain skeletal bond length which for polyethylenes is equal to 0.154nm [16].

Then the molecular weight of the chain part between nanoclusters M_{cl} was determined according to the equation [8]:

$$M_{cl} = \frac{\rho_p N_A}{v_{cl}}$$ (3.6),

where, ρ_p is polymer density, which for the studied polyethylenes is equal to $\sim 930 kg/m^3$, N_A is Avogadro number.

At last, the value n_{cl} is determined as follows [8]:

$$n_{cl} = \frac{2M_e}{M_{cl}}$$ (3.7),

where, M_e is molecular weight of a chain part between entanglements traditional nodes ("binary hookings"), which is equal to 1,390g/mole for low density polyethylenes [17].

In Figure 3.1, the dependence of nanocomposite elasticity modulus E_n on value n_{cl} is adduced, from which E_n enhancement at n_{cl} decreasing follows. Such behavior of nanocomposites LLDPE/MMT is completely identical to the behavior of both particulate-filled [11] and natural [10] nanocomposites.

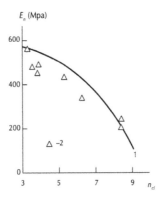

FIGURE 3.1 The dependences of elasticity modulus E_n on statistical segments number per nanocluster n_{cl} for nanocomposites LLDPE or MMT. *1* calculation according to the equation (3.8), *2* the experimental data.

In paper [18], the theoretical dependences of E_n as a function of cluster model parameters for natural nanocomposites were obtained:

$$E_n = c \left(\frac{\phi_{cl} V_{cl}}{n_{cl}} \right)$$ (3.8),

where, c is constant accepted equal to $5.9 \times 10^{-26} m^3$ for LLDPE.

In Figure 3.1, the theoretical dependence $E_n(n_{cl})$ calculated according to the equation (3.8) for the studied nanocomposites is adduced, which shows a good enough correspondence with the experiment (the average discrepancy of theory and experiment makes up 11.6% that is comparable with mechanical tests experimental error). Therefore, at organoclay mass contents W_n increasing within the range of 0–7mass%, n_{cl} value reduces from 8.40 to 3.17, that is, accompanied by nanocomposites LLDPE or MMT elasticity modulus growth from 206 to 569MPa.

Let us consider the physical foundations of n_{cl} reduction at W_n growth. The main equation of the reinforcement percolation model is the following one [9]:

$$\frac{E_n}{E_m} = 1 + 11 \left(\phi_n + \phi_{if} \right)^{1.7}$$ (3.9),

where, ϕ_n and ϕ_{if} are relative volume fractions of nanofiller and interfacial regions, accordingly.

The value ϕ_n can be determined according to the equation [5]:

$$\phi_n = \frac{W_n}{\rho_n}$$ (3.10),

where, ρ_n is nanofiller density which is equal to ~1,700kg/m³ for Na⁺-montmorillonite [5].

Further, the equation (3.9) allows to estimate the value ϕ_{if}. In Figure 3.2, the dependence $n_{cl}(\phi_{if})$ for nanocomposites LLDPE or MMT is adduced. As one can see, n_{cl} reduction at ϕ_{if} increasing is observed, that is, formed on organoclay surface densely-packed (and possibly, subjecting to epitaxial crystallization [9]) interfacial regions as if pull apart nanoclusters, reducing statistical segments number in them. As it follows from the equations (3.9) and (3.8), these processes have the same direction, namely, nanocomposite elasticity modulus increase.

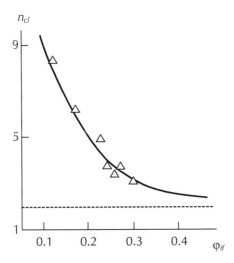

FIGURE 3.2 The dependence of statistical segments number per nanocluster n_{cl} on interfacial regions relative fraction φ_{if} for nanocomposites LLDPE/MMT. Horizontal shaded line indicates the minimum value $n_{cl} = 2$.

3.1.4 CONCLUSIONS

Hence, the results obtained in present paper demonstrated common reinforcement mechanism of natural and artificial (filled with inorganic nanofiller) polymer nanocomposites. The statistical segments number per one nanocluster reduction at nanofiller contents growth is such a mechanism on suprasegmental level. The indicated effect physical foundation is the densely-packed interfacial regions formation in artificial nanocomposites.

KEYWORDS

- **Linear low density polyethylene**
- **Nanocluster**
- **Na+-montmorillonite**
- **Organoclay strain**
- **Polymer Nanocomposites**

REFERENCES

1. Ahmed, S. & Jones, F. R. (1990). *Journal of Material Science, 25*(12), 4933–4942.
2. Bobryshev, A. N., Kozomazov, V. N., Babin, L. O., & Solomatov, V. I. (1994). *Synergetics of composite materials* (p. 154). Lipetsk: NPO ORIUS.
3. Kozlov, G. V., Yanovskii, Yu. G., & Zaikov, G. E. (2010). *Structure and properties of particulate-filled polymer composites: The fractal analysis* (p. 282). New York: Nova Science Publishers, Inc.
4. Hotta, S. & Paul, D. R. (2004). *Polymer, 45*(21), 7639–7654.
5. Sheng, N., Boyce, M. C., Parks, D. M., Rutledge, G. C., Abes, J. I., & Cohen, R. E. (2004). *Polymer, 45*(2), 487–506.
6. Bashorov, M. T., Kozlov, G. V., & Mikitaev, A. K. (2009). *Materialovedenie, 9,* 39–51.
7. Kozlov, G. V. & Novikov, V. U. (2001). *Uspekhi Fizicheskikh Nauk, 171*(7), 717–764.
8. Kozlov, G. V. & Zaikov, G. E. (2004). *Structure of the polymer amorphous state* (p. 465). Utretch: Brill Academic Publishers.
9. Mikitaev, A. K., Kozlov, G. V., & Zaikov, G. E. (2008). *Polymer nanocomposites: Variety of structural forms and applications* (p. 319). New York: Nova Science Publishers, Inc.
10. Kozlov, G. V. & Mikitaev, A. K. (2010). *Polymers as natural nanocomposites: Unrealized potential* (p. 323). Saarbrücken: Lambert Academic Publishing.
11. Edwards, D. C. (1990). *Journal of Material Science, 25*(12), 4175–4185.
12. Balankin, A. S. (1991). *Synergetics of deformable body* (p. 404). Moscow: Publishers of Ministry Defence SSSR.
13. Kozlov, G. V. & Sanditov, D. S. (1994). *Anharmonic effects and physical–mechanical properties of polymers* (p. 261). Novosibirsk: Nauka.
14. Budtov, V. P. (1992). *Physical chemistry of polymer solutions* (p. 384). St. Petersburg: Khimiya.
15. Aharoni, S. M. (1985). *Macromolecules, 18*(12), 2624–2630.
16. Aharoni, S. M. (1983). *Macromolecules, 16*(9), 1722–1728.
17. Wu, S. (1989). *Journal of Polymer Science Part B: Polymer Physics, 27*(4), 723–741.
18. Kozlov, G. V. (2011). *Recent Patents on Chemical Engineering, 4*(1), 53–77.

CHAPTER 4

PRODUCTION AND APPLICATION OF POLYMER NANOCOMPOSITES: PART II

G. V. KOZLOV, Z. M. ZHIRIKOVA, V. Z. ALOEV,
BOB A. HOWELL, and G. E ZAIKOV

CONTENTS

4.1 The Crystallization Kinetics of Nanocomposites Polypropylene or Carbon Nanotubes.. 44

    4.1.1 Introduction.. 44

    4.1.2 Experimental... 44

    4.1.3 Results and Discussion 45

    4.1.4 Conclusions.. 51

Keywords 51

References... 51

4.1   THE CRYSTALLIZATION KINETICS OF NANOCOMPOSITES POLYPROPYLENE OR CARBON NANOTUBES

4.1.1   INTRODUCTION

At present, strong enough and diverse changes of polymers crystalline structure occurring at the introduction in them of all sorts of fillers are well-known [1]. As a rule, these changes are described within the frameworks of polymer's crystalline morphology. However, lately the fractal model has been developed, which takes into consideration the whole complexity of polymers structure. It is assumed that changes occur not only on supramolecular level, but also on molecular and topological levels [2]. It is necessary to take into consideration simultaneously, that crystalline phase morphology variation causes non crystalline regions structure changes [3]. The introduction in semicrystalline polymer inorganic nanofiller results, as a rule, in polymer matrix crystallinity degree increase, since nanofiller particles serve as nucleators. Such effect was observed in nanocomposites high density polyethylene or calcium carbonate (HDPE or $CaCO_3$) [4, 5]. Let us note an important feature of semicrystalline polymers filling: nanofiller introduction can result in both reduction and enhancement of polymer matrix crystallinity degree K. So, the authors [6] found K decrease from 0.72 to 0.38 at carbon fibers introduction in HDPE at its volume content $\varphi_n = 0.303$. Therefore, the purpose of present paper is quantitative description of polypropylene crystalline phase structural changes on the above indicated structural levels after the introduction of carbon nanotubes in it.

4.1.2   EXPERIMENTAL

Polypropylene (PP) "Kaplen" of mark 01030 was used as a matrix polymer. This PP mark has melt flow index of 2.3–3.6g/10min, molecular weight of $\sim(2\text{--}3) \times 10^5$ and polydispersity index of 4.5.

Carbon nanotubes (CNT) of mark "Taunite", having an external diameter of 20–70nm, an internal diameter of 5–10nm and length of 2mcm and more, were used as nanofiller. They were prepared by chemical deposition (catalytic pyrolysis) of hydrocarbons (C_nH_m) on catalysts (Ni/Mg) at atmospheric pressure and temperature of 853–923K. CNT preparation process duration made up 10–80min. In the studied nanocomposites, CNT contents were changed within limits of 0.25–3.0mass %.

Nanocomposites PP or CNT were prepared by components mixing in melt on twin screw extruder Thermo Haake, model Reomex RTW 25/42, production of German Federal Republic. Mixing was performed at temperature 463–503K and screw speed of 50rpm during 5min. Testing samples were obtained by casting under pressure method on a casting machine Test Samples Molding Apparate RR/TS MP of firm Ray-Ran (Taiwan) at temperature 503K and pressure 8MPa.

Uniaxial tension mechanical tests have been performed on the samples in the shape of two-sided spade with sizes according to GOST 112 62-80. The tests have been conducted on universal testing apparatus Gotech Testing Machine CT-TCS 2000, production of German Federal Republic, at temperature 293K and strain rate $\sim 2 \times 10^{-3} s^{-1}$.

The nanocomposites PP or CNT crystallization kinetics was studied by differential scanning calorimetry (DSC) method on apparatus DSC 204 F1 Phoenix of the firm NETZSCH at scanning rate 10K/min. During the entire scanning time, the samples were in helium atmosphere with cleaning rate 25ml/min. The melting temperature was determined by DSC peaks, the greatest intensity position and crystallinity degree—by area under these peaks.

4.1.3   RESULTS AND DISCUSSION

A polymeric materials crystallinity degree K change is closely connected with kinetics of their crystallization, which can be described quantitatively according to the well-known Kolmogorov–Avrami equation [7]:

$$K = 1 - e^{-zt^n}$$

(4.1),

where, z is crystallization rate constant, t is crystallization process duration, n is Kolmogorov–Avrami exponent, characterizing nucleation and growing crystalline structures type for the given polymer material.

As it has been shown in paper [8], the exponent n is connected with fractal dimension D_{ch} of the chain part between local order domains (nanoclusters) as follows:

$$n = 3(D_{ch} - 1) + 1 \qquad (4.2).$$

The value D_{ch}, which characterizes molecular mobility level of polymeric material, can be calculated with the aid of following equation [9]:

$$\frac{2}{\phi_{cl}} = C_{\infty}^{D_{ch}} \qquad (4.3),$$

where, ϕ_{cl} is nanocluster's relative fraction, C_{∞} is characteristic ratio, which is polymer chain statistical flexibility indicator [10], and its value estimation method will be given below.

The value ϕ_{cl} can be calculated as follows. At first, the nanocomposite structure fractal dimension d_f is determined according to the equation [11]:

$$d_f = (d - 1)(1 + v) \qquad (4.4),$$

where, d is dimension of Euclidean space, in which a fractal is considered (it is obvious that in our case $d = 3$), v is Poisson's ratio, which is estimated according to the mechanical tests results with the aid of relationship [12]:

$$\frac{\sigma_Y}{E_n} = \frac{1 - 2v}{6(1 + v)} \qquad (4.5),$$

where, σ_Y is yield stress, E_n is nanocomposite elasticity modulus.

Further, the value C_{∞} can be determined according to the formula [9]:

$$C_{\infty} = \frac{2d_f}{d(d - 1)(d - d_f)} + \frac{4}{3} \qquad (4.6).$$

In Figure 4.1, the dependence $K(n)$ for nanocomposites PP or CNT is adduced. As it was expected, increase in K growth at n is observed. This dependence is described according to the following empirical approximation:

$$K = 0.8 \ (n - 1) \tag{4.7}.$$

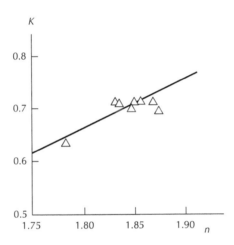

FIGURE 4.1   The dependence of crystallinity degree K on Kolmogorov-Avrami exponent n for nanocomposites PP or CNT.

As it follows from the Equation (4.7), the value $K = 0$ is reached at $n = 1.0$, or, according to the Equation (4.2), at $D_{ch} = 1.0$, that is, at fully suppressed molecular mobility. The value $K = 1.0$ is reached at $n = 2.20$ or $D_{ch} = 1.40$.

Since carbon nanotubes are simultaneously nucleator in nanocomposite matrix crystallization process, then with the increase in exponent n at CNT contents, φ_n growth should be expected. The nanofiller volume contents φ_n were calculated according to the known relationship [13]:

$$\phi_n = \frac{W_n}{\rho_n} \tag{4.8},$$

where, W_n is nanofiller mass content, ρ_n is its density, which is estimated as follows [13]:

$$\rho_n = 188\left(D_n\right)^{1/3} \text{ kg/m}^3 \qquad (4.9),$$

where, D_n is CNT diameter, which is given in nanometers.

In Figure 4.2, the dependence $n(\varphi_n^{2/3})$, where the value $\varphi^{2/3}$ charac-terizes CNT surface total area, which shows linear growth of n at $\varphi_n^{2/3}$ (or φ_n), increases. At $\varphi_n = 0$ the dependence $n(\varphi_n^{2/3})$ is extrapolated to $n \approx 1.82$, that is, approximately equal to Kolmogorov–Avrami exponent for the initial PP and at $\varphi_n = 1.0$ is greatest for studied nanocomposite value $n \approx 2.23$. Let us note that for carbon fibers (CF) in paper [6], the opposite effect was obtained, namely, n decrease at φ_n growth. Such discrepancy is explained by different surface structure of the used filler—CF and CNT. If in the first case fibers have smooth surface with dimension $d_{surf} \approx 2.15$, then in the second one, carbon nanotubes possess very rough surface with dimension $d_{surf} \approx 2.73$ [13]. As Pfeifer has been shown [14], such filler surface structure difference defines the difference of conformations of ad-joining to surface macromolecular coils—they are stretched (straightened) on smooth surface and maintain the initial conformation of a statistical coil on a rough one. In its turn, this defines D_{ch} decrease in the first case and constant value or this dimension enhancement—in the second one. For CNT, the dependence of n on their contents φ_n can be described by the following empirical equation:

$$n = n_{PP} + 2.90\phi_n \qquad (4.10),$$

where, n_{PP} is n value for the initial PP, which is equal to 1.78.

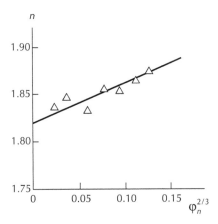

FIGURE 4.2   The dependence of Kolmogorov–Avrami exponent n on nanofiller contents φ_n for nanocomposites PP or CNT.

As it is known [7], the exponent n defines the forming crystalline phase morphology of polymeric materials. In the athermic nucleation case at $n \leq 2$, a ribbons are formed by two-dimensional growth mechanism, at $n \leq 3$—circles and at $n > 3$—spheres. Fractional values n means the combined mechanism of thermal or athermal nucleation, moreover, fractional part decrease indicates athermic mechanism role enhancement, that is, intensification of all crystallites growth simultaneous start [7]. As it follows from Figure 4.2 data, the exponent n, fractional part increasing at φ_n growth is observed. This means nucleation thermal mechanism role enhancement, that is, crystalline regions on carbon nanotubes growth intensification, besides, for the studied nanocomposites, the limiting value $n \approx 2.25$ assumes that in them spherical crystalline structures (spherolites) formation is impossible [7].

The theoretical value of crystallinity degree K^T can be determined according to the equation [2]:

$$K^T = 0.32C_\infty^{1/3} \tag{4.11}.$$

In Figure 4.3, the comparison of experimental K and calculated crystallinity degree values for nanocomposites PP or CNT according to the Equation (4.11), is adduced. As one can see, that the K^T values obtained by the indicated method are systematically lower than K. It is supposed that this effect is due to interfacial regions crystallization in the studied nanocomposites. The interfacial regions relative fraction φ_{if} can be estimated with the aid of the following equation [13]:

$$\frac{E_n}{E_m} = 1 + 11\left(\phi_n + \phi_{if}\right)^{1.7}$$ (4.12),

where, E_n and E_m are elasticity moduli of nanocomposite and matrix polymer, accordingly.

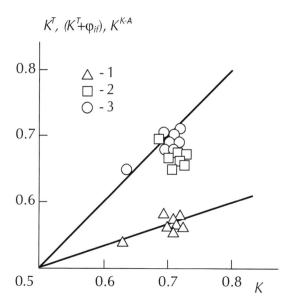

FIGURE 4.3 The comparison of experimental K, calculated according to the Equations (4.11). K^T (4.1), (4.11) and (4.12); $(K^T+\varphi_{if})$ (4.2) and (4.1); K^{K-A} (4.3) crystallinity degree values for nanocomposites PP or CNT.

In Figure 4.3, the comparison of the parameters K and (K^T+_{if}) for nanocomposites PP or CNT is adduced, which has shown their good enough correspondence (the average discrepancy of the indicated parameters makes up 6.2%). Hence, this observation assumes interfacial regions crystallization in the studied nanocomposites.

Besides the theoretical calculation of crystallinity degree K^{K-A} according to the Equation (4.1) performed at the following values with including in it parameters: $z = 0.06$ and $t = 5$. As it follows from Figure 4.3 data, in this case theory and experiment, excellent correspondence was obtained (the average discrepancy of K and K^{K-A} makes up 2.5% only).

4.1.4   CONCLUSIONS

Therefore, the present paper results have demonstrated molecular mobility decisive influence on crystallinity degree value for nanocomposites with semicrystalline polymer matrix. In its turn, fractal dimension of a chain part between nanoclusters, characterizing the indicated mobility, is defined by nanofiller surface structure. The offered fractal model of crystallization process is a universal one for polymer composites with semicrystalline matrix irrespective of filler type [15].

KEYWORDS

- **Carbon nanotubes (CNT)**
- **Crystallization**
- **Differential scanning calorimetry (DSC)**
- **Polypropylene (PP)**
- **Polymer Nanocomposites**

REFERENCES

1. Solomko, V. P. (1980). *The filled crystallizing polymers* (p. 264). Kiev: Naukova Dumka.
2. Aloev V. Z., & Kozlov, G. V. (2002). *The physics of orientational phenomena in polymeric materials* (p. 288). Nal'chik: Poligrafservis i T.

3. Kozlov, G. V., & Zaikov, G. E. (2003). *Izvestiya KBNC RAN, 1*(9), 54–57.
4. Tanniru, M., & Misra, R. D. K. (2005). *Material Science Engineering, 405A*(1), 178–193.
5. Deshmane, C., Yuan, Q., & Misra, R. D. K. (2007). *Material Science Engineering, 452–453A*(3), 592–601.
6. Dolbin, I. V., Burya, A. I., & Kozlov, G. V. (2008). *Kompozitsionnye Materialy, 2*(1), 3–7.
7. Wunderlich, B. (1976). *Macromolecular physics. Crystal nucleation, growth, annealing* (Vol. 2, p. 561). New York: Academic Press.
8. Aloev, V. Z., Kozlov, G. V., & Zaikov, G. E. (2001). *Russian Polymer News, 6*(4), 63–66.
9. Kozlov, G. V. & Zaikov, G. E. (2004). *Structure of the polymer amorphous state* (p. 465). Utrecht: Brill Academic Publishers.
10. Budtov, V. P. (1992). Physical chemistry of polymer solutions (p. 384). St. Petersburg: Khimiya.
11. Balankin, A. S. (1991). *Synergetics of deformable body* (p. 404). Moscow: Publishers of Ministry Defence SSSR.
12. Kozlov, G. V., & Sanditov, D. S. (1994). *Anharmonic effects and physical–mechanical properties of polymers* (p. 261). Novosibirsk: Nauka.
13. Mikitaev, A. K., Kozlov, G. V., & Zaikov, G. E. (2008). *Polymer nanocomposites: Variety of structural forms and applications* (p. 319). New York: Science Publishers, Inc.
14. Pfeifer, P. (1986). In L. Pietronero & E. Tosatti (Eds.), *Fractals in Physics* (pp. 70–79). Amsterdam: Elsevier.
15. Aphashagova, Z. Kh., Kozlov, G. V., & Malamatov, A. Kh. (2009). *Nanotekhnologii Nauka i Proizvodstvo, 1,* 40–44.

CHAPTER 5

COMPUTATION MODELING OF NANOCOMPOSITS ACTION ON THE DIFFERENT MEDIA

V. I. KODOLOV, N. V. KHOKHRIAKOV, V. V. TRINEEVA,
M. A. CHASHKIN, L. F. AKHMETSHINA, YU. V. PERSHIN, and
YA. A. POLYOTOV

CONTENTS

5.1 The Quantum Chemical Investigation of Metal / Carbon Nanocomposites Interaction With Different Media and Polymeric Compositions 55

    5.1.1 The Quantum Chemical Investigation of Hydroxyl Fullerene Interaction with Water 55

    5.1.2 Influence Comparison of Hydroxyl Fullerenes and Metal / Carbon Nanoparticles with Hydroxyl Groups on Water Structure 62

    5.1.3 Quantum-Chemical Investigation of Interaction Between Fragments of Metal–Carbon Nanocomposites and Polymeric Materials Filled by Metal Containing Phase 77

5.2 The Metal or Carbon Nanocomposites Influence Mechanisms on Media and on Compositions 85

5.3 The Compositions Modification Processes By Metal / Carbon Nanocomposites 93

    5.3.1 General Fundamentals of Polymeric Materials Modification by Metal / Carbon Nanocomposites 94

5.3.2   About the Modification of Foam Concrete by Metal or
        Carbon Nanocomposites Suspension................................... 97

5.3.3   The Modification of Epoxy Resins by Metal or Carbon
        Nanocomposites Suspensions... 100

5.3.4   The Modification of Glues Based on Phenol-Formaldehyde
        Resins by Metal or Carbon Nanocomposites Finely
        Dispersed Suspension.. 102

5.3.5   The Modification of Polycarbonate by
        Metal/Carbon Nanocomposites 104

Keywords 110

References...110

5.1   THE QUANTUM CHEMICAL INVESTIGATION OF METAL / CARBON NANOCOMPOSITES INTERACTION WITH DIFFERENT MEDIA AND POLYMERIC COMPOSITIONS

5.1.1   THE QUANTUM CHEMICAL INVESTIGATION OF HYDROXYL FULLERENE INTERACTION WITH WATER

Currently various sources contain experimental data that prove radical changes in the structure of water and other polar liquids when super small quantities of surface active nanoparticles (according to a number of papers—about 1,000th of percent by weight) are introduced. When the water modified with nanoparticles is further applied in technological processes, this frequently results in qualitative changes in the properties of products, including the improvement of various mechanical characteristics.

The investigation demonstrates the results of quantum-chemical modeling of similar systems that allow making the conclusion on the degree of nanoparticle influence on the structure of water solutions. The calculations were carried out in the frameworks of ab initio Hartree–Fock method in different basis sets and semi-empirical method PM3. The program complex GAMESS was applied in the calculations [1]. The equilibrium atomic geometries of fragments were defined and the interaction energy of molecular structures E_{int} was evaluated in the frameworks of above energy models. The interaction energy was calculated by the following formula:

$$E_{int} = E - E_a - E_b$$

(5.1)

where, E—energy of the molecular complex with an optimized structure, E_a and E_b—bond energies of isolated molecules forming the complex.

When calculating E_a and E_b, the molecule energy was not additionally optimized.

At the first stage, quantum-chemical investigation of ethyl alcohol molecule interaction with one and few water molecules was carried out. The calculations were carried out in the frameworks of different semi-empirical and *ab initio* models. Thus based on the data obtained, we can make a conclusion on the error value that can occur due to the use of simplified models with small basis sets. The calculation results are given in Table 5.1.

TABLE 5.1   Equilibrium parameters of the complex formed by the molecules of ethyl alcohol and water.

Calculation method	E_{int}(kcal/mol)	R_{9-10}	R_{3-9}	Q_3	Q_9	Q_{10}
PM3	1.88	2.38	0.95	-0.33	0.20	-0.37
3-21G	11.29	1.80	0.97	-0.74	0.41	-0.72
6-31G	7.52	1.89	0.96	-0.81	0.48	-0.83
TZV	6.90	1.91	0.96	-0.64	0.42	-0.83
3-21G MP2	13.17	1.78	1.00	-0.64	0.35	-0.63
TZV**++ MP2	6.27	1.90	0.97	-0.49	0.33	-0.65
TZV MP2	7.52	1.84	0.99	-0.58	0.39	-0.79
TZV+ MP2 for the geometry obtained by the method 6-31G	6.27	1.89	0.96			

The geometrical structure of the complex formed by the ethyl alcohol and water molecules are given in Figure 5.1.

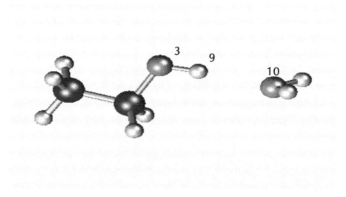

FIGURE 5.1   Equilibrium geometric structure of the complex formed by the molecules of ethyl alcohol and water. Optimization by energy is carried out in the basis 6-31G.

The energy minimization of the structure demonstrated was carried out on the basis 6-31G. The atom numbers in the Figure correspond to the designations given in the tables.

The calculations demonstrate that equilibrium geometrical parameters of the complex change insignificantly in the calculations by different methods. It should be noted that in the structures obtained without electron correlation the atoms 3, 9, 10, and hydrogen of water molecule are in one plane. The structure planarity is distorted when considering electron correlation by the perturbation theory MP2. Even more considerable deviations from the planarity are observed when calculating by the semi-empirical method PM3. In this case, the straight line connecting the atoms 3 and 9 become practically perpendicular to the water molecule plane. At the same time, the semi-empirical method gives the hydrogen bond length exceeding *ab initio* results by over 20%.

The analysis of energy parameters indicates that the energy of interaction of molecules inside the complex calculated by the Formula 5.1 is adequately predicted in the frameworks of Hartree–Fock method in the basis 6-31G. If for the geometrical structure obtained by this method the calculations of energy are made in the frameworks of more accurate models, the calculation results practically coincide with the successive accurate calculations.

The atom charges obtained by different methods vary in a broad range, in this case no regularity is observed. This can be connected with the imperfection of the methodology of charge evaluation by Mulliken.

The investigation of water interaction with the cluster of hydroxyfullerene $C_{60}[OH]_{10}$ was carried out in the frameworks of *ab initio* Hartree–Fock method in the basis 6-31G. To simplify the calculation model the cluster with the rotation axis of 5th order of fullerene molecule C_{60} was applied. Figure 5.2a demonstrates the fullerene molecule structure optimized by energy, Figure 5.2b—the optimized structure of the cluster $C_{60}[OH]_{10}$.

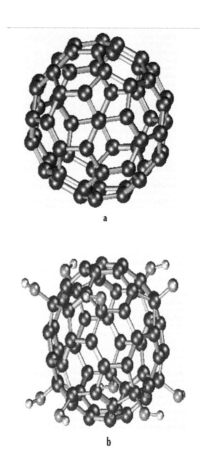

FIGURE 5.2 Equilibrium geometric structure of the molecule C_{60} (a) and cluster $C_{60}[OH]_{10}$ (b). Optimization by energy is carried out in the basis 6-31G.

All the calculations were carried out preserving the symmetry $C_5 C_5$ of molecular systems.

The pentagon side length calculated in the fullerene molecule is 1.452Å, and bond length connecting neighboring pentagons is 1.375Å. These results fit the experimental data available and results of other calculations [2].

To evaluate the energy of interaction of the cluster $C_{60}[OH]_{10}$ with water, the complex $C_{60}[OH]_{10} \cdot 10 \ H_2O$ with the symmetry C_5 was studied. The complex geometry obtained in the basis 6-31G is given in Figure 5.3.

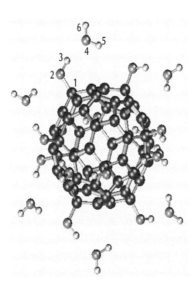

FIGURE 5.3 Molecular system formed by the cluster $C_{60}[OH]_{10}$ and 10 water molecules. Optimization by energy is carried out in the basis 6-31G.

The complex geometrical parameters and atom charges are shown in Table 5.2.

TABLE 5.2 Equilibrium parameters of the complex $C_{60}[OH]_{10} \cdot 10\,H_2O$ (calculations by *ab initio* method in the basis 6-31 G).

Charges					
Q_1	Q_2	Q_3	Q_4	Q_5	Q_6
0.17	-0.78	0.50	-0.86	0.46	0.42

Bond lengths				
R_{1-2}	R_{2-3}	R_{3-4}	R_{4-5}	R_{4-6}
1.42	0.97	1.86	0.95	0.95

The atom numeration in the table corresponds to the numeration given in Figure 5.3.

TABLE 5.3 Characteristics of the complexes formed by the molecules $C_{60}[OH]_{10}$, C_2H_5OH with water. *Ab initio* calculations in the basis 6-31G.

	E_{int} (kcal/mol)	L,A	Q_H	Q_0 (OH)	Q_0 (H$_2$O)
$C_{60}(OH)_{10}$ + 10H$_2$O	12.69	1.86	0.5	-0.78	-0.86
C_2H_5OH + H$_2$O	7.52	1.89	0.47	-0.81	-0.83
H$_2$O + H$_2$O	8.37	1.89	0.42	-0.82	-0.76

Table 5.3 contains the comparison of the breaking off energy of water molecule from the complexes $C_{60}[OH]_{10} \cdot 10\,H_2O$, $C_2H_5OH \cdot H_2O$, $H_2O \cdot H_2O$. Besides, the table gives the lengths of hydrogen bonds in these complexes L and charges of Q atoms of oxygen and hydrogen of OH group and oxygen atom of water molecule. All the calculations are carried out by *ab initio* Hartree–Fock method in the basis 6-31G. The calculations demonstrate that the energy of interaction of ethyl alcohol molecule with water is somewhat less than the energy of interaction between water molecules. At the same time, nanoparticle $C_{60}[OH]_{10}$ interacts with water molecule about two times more intensively. When comparing the atom charges of the complexes it can be seen that the molecule π—fullerene electrons in the nanoparticle act as a reservoir for electrons. Thus the molecule OH group is practically electrically neutral. This intensifies the attraction of water molecule oxygen. Therefore the strength of hydrogen bond in the complex $C_{60}[OH]_{10} \cdot 10\,H_2O$ increases.

To evaluate the cluster sizes with the center in nanoparticle $C_{60}[OH]_{10}$ that can be formed in water, semi-empirical calculations of energies of attachment of water molecules to the nanoparticle $C_{60}[OH]_2$ with the chain formation are carried out (Figure 5.4). Semi-empirical method PM3 was applied in the calculations. Water molecules are attached to the chain successively; after each molecule is attached, the energy is minimized.

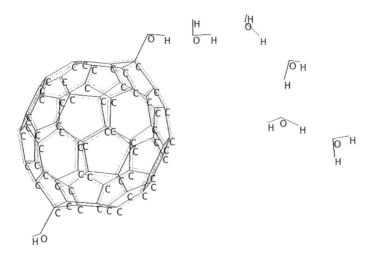

FIGURE 5.4   Model system formed by the cluster $C_{60}[OH]_{10}$ and chain from 5 water molecules. Optimization by energy is carried out by semi-empirical method PM3.

Table 5.4 contains the results obtained for the chain growth from the nanoparticle, alcohol molecule, and water molecule. Bond energies of water molecules with the chain growing from $C_{60}[OH]_2$ over two times exceed similar energies for the chain growing from OH-group of alcohol molecule and water molecule.

TABLE 5.4   Energies of attachment of water molecule to the chain connected with the molecules $C_{60}[OH]_{10}$, C_2H_5OH. Semi-empirical calculations by the method PM3.

n	$(H_2O)_n$	$C_2H_5OH + nH_2O$	$C_{60}(OH)_2 + nH_2O$
1		-1.40	-4.06
2	-1.99	-2.19	-4.83
3	-2.38	-2.51	-5.02
4	-2.54	-2.58	-4.88
5	-2.56	-2.68	-4.71
6	-2.66		-3.04

At the same time, the energy of hydrogen bond first increases till the attachment of the third molecule, and then gradually decreases. When the sixth molecule is attached, the energy decreases in discrete steps.

Thus the investigations demonstrate that hydroxyfullerene molecule forms a stable complex in water, being surrounded by six layers of water molecules. Obviously the water structure is reconstructed at considerably larger distances. The introduction of insignificant number of nanoparticles results in complete reconstruction of water medium, and, therefore, in considerable change in the properties of substances obtained on its basis. Thus, the results are reported on the considerable change in the properties of constructional materials obtained on the water basis after the insignificant additions of active graphite-like nanoparticles with chemical properties similar to the ones of hydroxyfullerene.

5.1.2 INFLUENCE COMPARISON OF HYDROXYL FULLERENES AND METAL / CARBON NANOPARTICLES WITH HYDROXYL GROUPS ON WATER STRUCTURE

Quantum-chemical investigations of the interactions between hydroxylated graphite-like carbon nanoparticles, including metal containing ones, and water molecules are given. Hydroxyfullerene molecules, as well as carbon envelopes consisting of hexangular aromatic rings and envelope containing a defective carbon ring (pentagon or heptagon) are considered as the models of carbon hydroxylated nanoparticles. The calculations demonstrate that hydroxyfullerene molecule $C_{60}[OH]_{10} \cdot (H_2O)_{10}$ forms a stronger hydrogen bond with water molecule than the bond between water molecules. Among the unclosed hydroxylated carbon envelopes, the envelope with a pentagon forms the strongest hydrogen bond with water molecule. The presence of metal nanocluster increases the hydrogen bond energy in 1.5–2.5 times for envelopes consisting of pentagons and hexagons.

At present, we can find in the literature experimental data confirming radical changes in the structure of water and other polar liquids when small quantities (about thousandth percent by mass) of surfacants are introduced [3–5]. When further used in technological processes, the water modified by nanoparticles frequently results in qualitative changes of product properties including the improvement of different mechanical characteristics.

The paper is dedicated to the investigation of water molecule interactions with different systems modeling carbon enveloped nanoparticles, including those containing clusters of transition metals. All the calculations were carried out following the same scheme. At the first stage the geometry of complete molecular complex was optimized. At the second stage the interaction energy of the constituent molecular structures E_{int} was assessed for all optimized molecular complexes. The interaction energy was calculated by the formula $E_{int} = E - E_a - E_b$, where E —bond energy of the optimized structure molecular complex, E_a and E_b—bond energies of isolated molecules forming the complex. When calculating E_a and E_b, the molecule energy was not additionally optimized. The energy of chemical bonds was assessed by the same scheme. All the calculations were carried out with the help of softwarePC-GAMESS[6].Bothsemi-empiricaland*ab initio*methodsofquantum chemistry with different basis sets were applied in the calculations. Besides, the method of density functional was used. The more detailed descriptions of calculation techniques are given when describing the corresponding calculation experiments.

At the first stage, the quantum-chemical investigation of the interaction of α-naphthol and β-naphthol molecules with water was carried out within various quantum-chemical models. Thus, based on the data obtained the conclusion on the error value can be made, which can be resulted from the application of simplified methods with insufficient basis sets. There are three isomers of naphthol molecule named α-naphthol (complex with water is shown on Figure 5.5a, trans-rotamer and cis-rotamer of β-naphthol (Figure 5.5b, c).

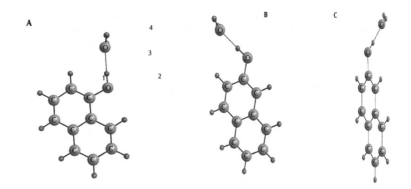

FIGURE 5.5 Equilibrium geometric structure of the complex formed by the molecules of α-naphthol (a), trans-β-naphthol (b), and cis-β-naphthol (c) with water. The optimization by energy was carried out at HF/6-31G level.

In what follows we denote them as α-naphthol, trans-β-naphthol and cis-β-naphthol correspondingly. The calculations within all methods applied, except for the semi-empirical one, showed that cis-β-naphthol is the most stable isomer, α-naphthol has almost equal stability, and trans-β-naphthol is the least stable structure. This conclusion is in accordance with available experimental data [7, 8].

The results for the complexes with α-naphthol are given in the Table 5.5.

Table 5.5 Equilibrium parameters of the complex formed by α-naphthol molecules and water. In the column E_{int} the interaction energy between water molecules in dimer is given in brackets for comparison.

Calculation method	E_{int} (kJ/mol)	r_{1-2} (nm)	r_{2-3} (nm)	r_{2-4} (nm)	q_1	q_2	q_3
PM3	16.27(8.32)	0.1364	0.0963	0.1819			
HF 3-21G	68.66(46.11)	0.1361	0.098	0.1701	0.45	-0.82	0.44
HF 6-31G	43.91(33.13)	0.1365	0.0961	0.1805	0.39	-0.85	0.50
HF 6-31G*	31.91(23.61)	0.1345	0.0954	0.194	0.42	-0.81	0.52
HF 6-31G* MP2	42.65(30.75)	0.1367	0.0984	0.1855	0.31	-0.72	0.48
DFT 6-31G* B3LYP	42.58(32.33)	0.136	0.0983	0.1821	0.28	-0.69	0.44

The geometry structure of the complex formed by naphthols and water molecules is shown in Figure 5.5. The structures given in Figure 5.5 were obtained by minimizing the energy using Hartree–Fock method in the basis 6-31G. The numbers of atoms in Figure 5.5 correspond to those given in the table. The calculations demonstrate that equilibrium geometric parameters of the complex vary insignificantly when applying different calculation methods. It should be pointed out that in the structures obtained at HF/6-31G level of theory without taking into account of electron correlation, atoms 2, 3, 4, and hydrogens of water molecule are almost in one plane. The structure planarity is broken with the inclusion of polarization functions into the basis set and electron correlation consideration. The semi-empirical calculation forecasts the planarity disturbance of the atom group considered. The hydrogen bond for β-naphthol complexes is longer than one for α-naphthol. These values are 0.1812nm for cis-β-naphthol and 0.1819nm for trans-β-naphthol calculated at HF/6-31G level. The values obtained at MP2/6-31G* level of theory are 0.1861nm and 0.1862nm correspondingly. They are in accordance with the results of more accurate calculations 0.1849nm and 0.1850nm [9].

The second column in Table 5.1 demonstrates the energy of interaction between α-naphthol and water molecule calculated using the Equation 5.1. The interaction energies between water molecules in water dimer are given in brackets for comparison. Although all energy characteristics for different methods vary in a broad range, the ratio of energies E_{int} for the complex of water with α-naphthol and water dimer is between 1.3 and 1.4. Thus, we can assume that Hartree–Fock method in the basis 6-31G provides qualitatively correct conclusions for the energies of intermolecular interaction. It should be pointed out that energy E_{int} calculated in the basis 6-31G agrees numerically with the results of more accurate calculations. Nevertheless the interaction energies between water and α-naphthol exceed experimental value 24.2 ± 0.8kJ/mol [5]. The energies of interaction between water molecules in dimer are overestimated as well. The errors are known to be caused by the superposition of basis sets of complex molecules (BSSE) and zero-point vibration energy (ZPVE). Calculations completed have shown that taking into account BSSE decreases the interaction energy by 10–20% depending on the basis, but does not lead to qualitative changes. Taking into account ZPVE results in additional decrease in the energies of intermolecular interaction. Thus taking into account ZPVE and BSSE considerably improves the accordance with the available data. Interaction energy between water and α-naphthol is 30kJ/mol in the basis 6-31G after correction. In general, analysis of energy parameters shows that all the models considered give the correct correlation for energies and demonstrate that the energy of interaction between water and α-naphthol molecules exceeds the energy of interaction between water molecules. Numerical accordance with experimental value is achieved only after taking into account electronic correlation at the MP2 level. In this model the interaction energy after all the corrections mentioned becomes 23.7kJ/mol; Cis-β-naphthol is known to be more stable than trans-isomer. The energy difference calculated in 6-31G basis set is 3.74kJ/mol that is equal to experimental value [8]. HF/6-31G binding energy for water dimer is 18.9kJ/mol after corrections (experimental value is 15 ± 2kJ/mol [10]). As in the case of α-naphthol complex MP2 calculation is necessary to obtain value that is equal to experimental value.

The vibrational frequencies of naphthols and their complexes with water molecule were calculated and multiplied by a scaling factor of 0.9034 [8]. IR-active frequencies corresponding to naphthol OH-group stretching are 3,660cm^{-1} for trans-β-naphthol and 3,654cm^{-1} for cis-β-naphthol. They are in accordance with experimental frequencies 3,661 and 3,654cm^{-1} [8]. The influence of hydrogen bond on naphthol OH-group stretching calculated in the present article is overestimated. Calculated frequencies at HF/6-31G level of theory in the presence of water molecule are 3,479cm^{-1} for trans-β-naphthol and 3,466cm^{-1} for cis-β-naphthol (for comparing experimental values are 3,523 and 3,512cm^{-1} correspondingly). The same result was observed in research [8]. The atom charges obtained within different methods vary in a broad range without any regularity. It can be connected with the imperfectness of Mulliken's charge evaluation technique.

Thus comparison of calculation results with experimental data demonstrates *ab initio* Hartree–Fock method in the basis 6-31G basis set provides qualitatively correct conclusion on hydrogen bond energies for systems considered. This model was applied to investigate the interaction between water and hydroxyfullerene cluster $C_{60}[OH]_{10}$. Hydroxyfullerene is a proper object for quantum chemical modeling of interaction between activated carbon nanoparticle and water. At the same time hydroxyfullerene is of great interest because of its possible applications in medicine, for water disinfection and for polishing nanosurfaces.

The cluster with the rotation axis C_5 was used in our research of hydroxyfullerene to simplify the calculation model. Figure 5.6a demonstrates the structure of C_{60} molecule optimized by energy, Figure 5.6b—the optimized structure of cluster $C_{60}[H]_{10}$. All the calculations were carried out conserving the symmetry C_5 of molecular systems.

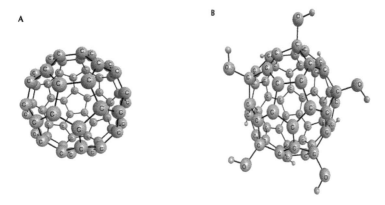

A   B

FIGURE 5.6   Equilibrium geometric structure of C_{60} molecule (a) and cluster $C_{60}[OH]_{10}$ (b). The optimization by energy is carried out at HF/6-31G level.

TABLE 5.6   Equilibrium parameters of the complex $C_{60}[OH]_{10} \cdot 10H_2O$ (HF/6-31G calculation).

Charges					
q_1	q_2	q_3	q_4	q_5	q_6
0.17	-0.78	0.50	-0.86	0.46	0.42

Bond lengths (nm)				
r_{1-2}	r_{2-3}	r_{3-4}	r_{4-5}	r_{5-6}
0.142	0.097	0.186	0.095	0.095

TABLE5.7 Characteristics of the complexes formed by $C_{60}[OH]_{10}$ and α-naphthol molecules and water. HF/6-31G calculation.

Complex	E_{int} (kJ/mol)	L (nm)	q_H	q_0 (OH)	q_0 (H$_2$O)
$C_{60}[OH]_{10} - 10H_2O$	53.09	0.186	0.5	-0.78	-0.86
$C_{10}H_7OH \cdot H_2O$	43.91	0.181	0.5	-0.85	-0.84
$H_2O. H_2O$	33.13	0.189	0.47	-0.88	-0.83

The length of pentagon side in the fullerene molecule was 0.1452nm, and the length of bond connecting the neighboring pentagons is 0.1375nm. These results agree with the experimental data available and results of other calculations [11, 12].

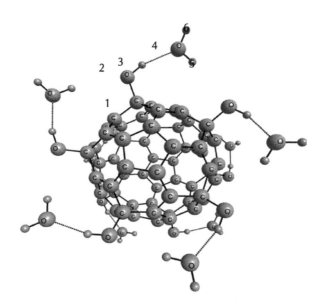

FIGURE 5.7   Molecular system formed by the cluster $C_{60}[OH]_{10}$ and 10 water molecules. The optimization by energy is carried out at HF/6-31G level.

Complex $C_{60}[OH]_{10} \cdot 10H_2O$ with symmetry C_5 was considered to obtain the interaction energy of cluster $C_{60}[OH]_{10}$ and water. The geometric structure of the complex optimized by energy is given in Figure 5.7. The complex geometric parameters and atom charges are given in Table 5.6. The atoms are enumerated in the same way as in Figure 5.7. Table 5.7 contains the water molecule isolation energies from complexes $C_{60}[OH]_{10}$ · $10H_2O$, $C_{10}H_7OH$ · H_2O, H_2O · H_2O. Besides, the table contains the lengths of hydrogen bonds in these complexes L and charges Q of oxygen and hydrogen atoms of fullerene OH-group and oxygen atom of water molecule. The calculations demonstrate that the energy of interaction between nanoparticle $C_{60}[OH]_{10}$ and water molecule nearly twice as big as the interaction energy of water molecules in dimer. Basis set superposition error was also analyzed. The calculation indicates that the account of this effect decreases the energies of intermolecular interaction by 15%. The qualitative conclusions from the calculations do not change; the gain in the interaction energy for the complex $C_{60}[OH]_{10}$ · $10H_2O$ increases insignificantly in comparison with other complexes. When comparing the atom charges of the complexes, it can be concluded that fullerene π-orbitals in nanoparticle act as a reservoir for electrons. Thus, the charge of hydroxyfullerene OH group decreases. This increases the attraction of water molecule oxygen and the strength of hydrogen bond in complex $C_{60}[OH]_{10}$ · $10H_2O$.

To estimate the size of the cluster with the center in nanoparticle $C_{60}[OH]_{10}$ that can be formed in water, the semi-empirical calculations of energies of water molecules adding to nanoparticle $C_{60}[OH]_{10}$ with chain formation were made (Figure 5.8). The semi-empirical method PM3 was used in calculations. Water molecules were bonded to the chain consequently; the energy was minimized after each molecule was added. Table 5.8 demonstrate the comparative results obtained for the chain growth both from the nanoparticle and water molecule. The binding energy of water molecule with the chain growing from $C_{60}[OH]_2$ over twice as big as similar energies for the chain growing from the water molecule. At the same time, the energy of hydrogen bond first increases before the addition of the third molecule, and then slowly decreases. When the sixth molecule is added, the energy decreases step-wise. Thus, the research demonstrates that hydroxyfullerene molecule forms a stable complex in water, being surrounded with six layers of water molecules. It is clear that the water structure changes at much longer distances. The introduction of insignificant number of nanoparticles results in complete reconstruction of aqueous medium, and, consequently, to sufficient changes in the properties of substances obtained on its base.

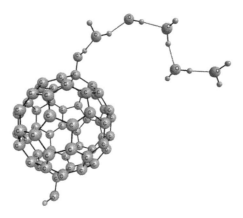

FIGURE 5.8   Model system formed by cluster $C_{60}[OH]_{10}$ and 5 water molecule chain. The optimization by energy was carried out with semi-empirical method PM3.

TABLE 5.8   Energies of binding water molecules to the chain connected with molecules $C_{60}[OH]_{10}$ and water chain (in kJ/mol). Semi-empirical calculations with PM3 method.

n	$[H_2O]_n$	$C_{60}[OH]_2 + n \cdot H_2O$
1		16.99
2	8.33	20.21
3	9.96	21.00
4	10.63	20.42
5	10.71	19.71
6	11.13	12.72

In some experimental works carbon nanoparticles are obtained in gels of carbon polymers in the presence of salts of transition metals [13]. At the same time, nanocomposites containing metal nanoparticles in carbon envelopes are formed. The envelope structure contains many defects but it mainly has an aromatic character. The nanoparticles considered are widely used for the modification of different substances; therefore, it is interesting to investigate the influence of metal nuclei on the interaction of carbon enveloped nanoparticles with polar fluids.

The systems demonstrated in Figures 5.9–5.11 were considered as the nanoparticle model containing a metal nucleus and hydroxylated carbon envelope. In all the figures of the paper carbon and hydrogen atoms are not signed. At the same time, carbon atoms are marked with dark spheres, and hydrogen atoms with light ones. Clusters comprising two metal atoms are introduced into the system to model a metal nucleus. Carbon envelopes are imitated with 8-nucleus polyaromatic carbon clusters. The carbon envelope indicated in Figure 5.9 contains one pentagonal aromatic ring, and in the envelope in Figure 5.11—one heptagon. The envelope in Figure 5.10 contains only hexagonal aromatic rings. The corresponding number of hydrogen atoms is located on the boundaries of carbon clusters. Besides, one OH-group is present in each of the carbon envelopes. Thus, the chemical formula of hydroxylated carbon envelope with pentagonal defect $C_{23}H_{11}OH$ hydroxylated with defectless carbon envelope—$C_{27}H_{13}OH$, and for the hydroxylated carbon envelope with heptagonal defect—$C_{31}H_{15}OH$. Copper and nickel are taken as metals. The interaction of the considered model systems with water molecules; the interaction of the considered model systems with water molecules is investigated in the paper.

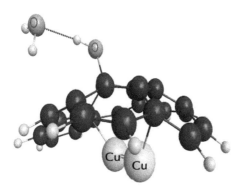

FIGURE 5.9   Complex $Cu_2C_{23}H_{11}OH.H_2O$ with hexagonal defect in carbon envelope.

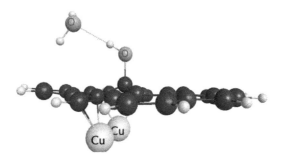

FIGURE 5.10   Complex $Cu_2C_{27}H_{13}OH.H_2O$.

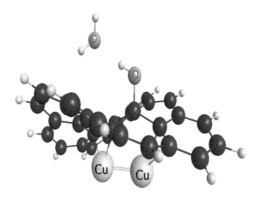

FIGURE 5.11   Complex $Cu_2C_{31}H_{15}OH.H_2O$ with heptagonal defect in carbon envelope.

The calculations are made in the frameworks of density functional with exchange-correlative functional B3LYP. For all the atoms of the model system, apart from oxygen, the basis set 6-31G was applied. Polarization and diffusion functions were added to the basis set for more accurate calculations of hydrogen bond energy and interaction of hydroxyl group with carbon envelope. Thus, the basis set 6-31+G* was used for for oxygen atoms. The basis sets from [14] were used for atoms of transition metals; the bases from software PC-GAMESS were applied for the rest of the elements.

Hydrogen bond energy in water dimers calculated in this model exceeded 22kJ/mol being a little more than the value obtained in the frameworks of the more accurate method—19.65kJ/mol [15]. The experimental value is 14.9 ± 2 kJ/mol [15]; if we consider the energy of zero oscillations of dimer and water molecules in the frameworks of the calculation model, hydrogen bond energy decreases to 12.57kJ/mol and becomes similar to the experimental value. When the basis extends due to polarization functions on hydrogen atoms, hydrogen bond energy goes down by 2%.

TABLE 5.9 Bond lengths (in angstrems).

Compound	$C-O$	$O-H$	$O \cdots H$
$C_{23}H_{11}OH$	1.492	0.973	–
$Cu_2C_{23}H_{11}OH$	1.450	0.973	–
$Ni_2C_{23}H_{11}OH$	1.454	0.973	–
$C_{27}H_{13}OH$	1.545	0.974	–
$Cu_2C_{27}H_{13}OH$	1.469	0.973	–
$Ni_2C_{27}H_{13}OH$	1.464	0.972	–
$C_{31}H_{15}OH$	1.506	0.973	–
$Cu_2C_{31}H_{15}OH$	1.470	0.973	–
$Ni_2C_{31}H_{15}OH$	1.455	0.972	–
H_2O	–	0.969	–
$C_{23}H_{11}OH \cdot H_2O$	1.470	0.981	1.967
$Cu_2C_{23}H_{11}OH \cdot H_2O$	1.439	0.984	1.869
$Ni_2C_{23}H_{11}OH \cdot H_2O$	1.440	0.985	1.893
$C_{27}H_{13}OH \cdot H_2O$	1.524	0.979	2.119
$Cu_2C_{27}H_{13}OH \cdot H_2O$	1.451	0.984	1.911

TABLE 5.9 *Continued*

$Ni_2C_{27}H_{13}OH \cdot H_2O$	1.449	0.981	1.964
$C_{31}H_{15}OH \cdot H_2O$	1.490	0.979	2.024
$Cu_2C_{31}H_{15}OH \cdot H_2O$	1.453	0.981	1.964
$Ni_2C_{31}H_{15}OH \cdot H_2O$	1.446	0.979	1.997
$H_2O \cdot H_2O$	–	0.978	1.906

At the first stage of the investigation, we minimized the energy of all the considered structures in the presence of water molecule and without it. In Figures 5.9–5.11 one can see the optimized geometric structures for complexes containing copper cluster. In the second column of Table 5.5 one can find the lengths of $C - O$ bond formed by OH-group with carbon envelope atom for the calculated complexes; in the third column, the bond length in hydroxyl group. In the fourth one, the lengths of hydrogen bond for complexes with a water molecule.

The analysis of geometric parameters demonstrates that in all cases a metal presence results in the decrease of $C - O$ bond length formed by OH-group with the carbon envelope atom. The especially strong effect is observed in the defectless carbon envelope which is explained by a high stability of defectless graphite-like systems. The metal availability results in the increased activity of the carbon envelope due to the additional electron density and destabilization of aromatic rings. The defects of carbon grid have a similar effect. At the same time, the pentagonal defect is usually characterized by the electron density excess and the heptagonal one by its lack. If we compare the two metals clusters considered, we see that for the system with pentagonal defect copper influence is stronger than nickel one, and for the rest of the systems we observe the opposite effect. A metal has only an insignificant influence on $O - H$ bond length.

The interaction with water molecule results in the increase of $O - H$ bond length in the hydroxyl group of carbon envelope and decrease of $C - O$ bond length. In the presence of metal the length of hydrogen bond decreases in all the cases, copper clusters cause the formation of shorter hydrogen bonds. In our calculations, we took into account the basis superpositional error (BSSE). In Table 5.6 one can see the interaction energies of the considered model systems with water molecule. In the table the energy of intermolecular interaction in water dimer calculated using the basis 6-31+G* is given in the table for comparison.

From the table, it is seen that the interaction of hydroxylated carbon envelopes with water molecule is significantly weaker than the intermolecular interaction in water dimer. The strongest hydrogen bond is formed in case of the envelope with pentagonal defect $C - O$. This result correlates with the minimal length of $C - O$ bond in the envelope $C_{23}H_{11}OH$ (Table 5.9). Thus, hydrogen bond is getting weaker in the hydroxyl group, and it provides the formation of a stronger hydrogen bond. The main conclusion from Table 5.6 is a considerable strengthening of hydrogen bond in metal systems. For complex $Ni_2C_{27}H_{13}OH \cdot H_2O$ with defectless carbon envelope the energy of hydrogen bond increases in 2.5 times in comparison with a similar complex without metal. The greatest energy of hydrogen bond is obtained for complex $Ni_2C_{27}H_{13}OH \cdot H_2O$. This energy exceeds the energy of hydrogen bond in water dimer by over 40%.

TABLE 5.10 Interaction energy of model clusters with water molecule (kJ/mol).

Compound	E_{int}
$C_{23}H_{11}OH \cdot H_2O$	18.12
$Cu_2C_{23}H_{11}OH \cdot H_2O$	26.06
$Ni_2C_{23}H_{11}OH \cdot H_2O$	31.56
$C_{27}H_{13}OH \cdot H_2O$	11.01
$Cu_2C_{27}H_{13}OH \cdot H_2O$	27.04
$Ni_2C_{27}H_{13}OH \cdot H_2O$	19.65
$C_{31}H_{15}OH \cdot H_2O$	15.08

TABLE 5.10 *Continued*

$Cu_2C_{31}H_{15}OH \cdot H_2O$	13.75
$Ni_2C_{31}H_{15}OH \cdot H_2O$	16.88
$C_{60}(OH)_2 \cdot H_2O$	21.98
$H_2O \cdot H_2O$	22.37

It should be noted that in the frameworks of model HF/6-31G the energy of hydrogen bond for hydroxyfullerene $C_{60}(OH)_{10} \cdot (H_2O)_{10}$ is higher than the energy of hydrogen bond in water dimer by 60% (Table 5.7). The semi-empirical calculations for complex $C_{60}(OH)_2 \cdot H_2O$ predict that the energy of hydrogen bond two times exceeds the one in water dimer. When calculating by the method described in the final part of this paper, the energy of hydrogen bond for hydroxyfullerene $C_{60}(OH)_{10} \cdot (H_2O)_{10}$ is higher by 37% than the bond energy in water dimer. Thus, taking into account diffusion and polarization functions we considerably change the results. The additional reason for the difference in the calculation of complex $C_{60}(OH)_{10} \cdot (H_2O)_{10}$ is the use of the method of hydrogen bond energy evaluation with breaking off all water molecules while preserving the system symmetry.

The main conclusion from the calculation results presented is a considerable increase in hydrogen bond energy with water molecule for carbon graphite-like nanoparticles containing the nucleus from transition metal atoms. Besides, the energy of C – 0 bond between carbon envelope and hydroxyl group increases in the presence of metal. Thus, the probability of the formation of hydroxylated nanoparticles increases. The nanoparticles considered have a strong orientating effect on water medium and become the embryos while forming materials with improved characteristics. A similar effect is observed when water molecule interacts with hydroxyfullerene. In this case, the interaction energy depends considerably on the number of OH-groups in the molecule. Thus, the interaction energy of water molecules with hydroxyfullerene in cluster $C_{60}(OH)_{10} \cdot (H_2O)_{10}$ exceeds the interaction energy in water dimer by 40–60% depending on the calculation model на 40–60%. At the same time, for complex $C_{60}(OH)_2 \cdot H_2O$ hydrogen bond is a little weaker than in water dimer.

5.1.3 QUANTUM-CHEMICAL INVESTIGATION OF INTERACTION BETWEEN FRAGMENTS OF METAL–CARBON NANOCOMPOSITES AND POLYMERIC MATERIALS FILLED BY METAL CONTAINING PHASE

There is a strong necessity to define formulations components processes conditions for nanostructured materials filled by metallic additives. Another task is optimization of components, nanocomposites and diluents combination and, in what follows, curing processes with determined temperature mode. The result of these arrangements will be materials with layer wise homogeneous metal particles /nanocomposites distribution formulation in ligand shell.

The result of these arrangements will be homogeneous metal particles / nanocomposites distribution in acetylacetone ligand shell. In the capacity of conductive filler silver nanoclusters or nanocrystals and copper-nickel / carbon nanocomposites can be used. It should be lead to decrease volume resistance from 10^{-4} to $10^{-6}\Omega$cm and increase of adhesive or paste adhesion.

Silver filler is more preferable due to excellent corrosion resistance and conductivity, but its high cost is serious disadvantage. Hence alternative conductive filler, notably nickel-carbon nanocomposite was chosen to further computational simulation. In developed adhesive or paste formulations metal containing phase distribution is determined by competitive coordination and cross-linking reactions.

There are two parallel technological paths that consist of preparing blend based on epoxy resin and preparing another one based on polyethylene polyamine (PEPA). Then prepared blends are mixed and cured with diluent gradual removing. Curing process can be tuned by complex diluent contains of acetiylacetone (AcAc), diacetonealcohol (DAcA) (in ratio AcAc:DAcA = 1:1) and silver particles being treated with this complex diluent. Epoxy resin with silver powder mixing can lead to homogeneous formulation formation. Blend based on PEPA formation comprises AA and copper- or nickel-carbon nanocomposites introducing. Complex diluent removing tunes by temperature increasing to 150–160°C. Initial results formulation with 0.01% nanocomposite (by introduced metal total weight) curing formation showed that silver particles were self-organized and assembled into layer-chained structures.

Acetylacetone (AcAc) infrared spectra were calculated with Hyper-Chem software. Calculations were carried out by the use of semi-empirical methods PM3 and ZINDO/1 and *ab initio* method with 6-31G** basis set [16]. Results of calculation have been compared with instrumental measurement performed IR-Fourier spectrometer FSM-1201. Liquid adhesive components were preliminary stirred in ratio 1:1 and 1:2 for the purpose of interaction investigation. Stirring was carried out with magnetic mixer. Silver powder was preliminary crumbled up in agate mortar. Then acetylacetone was added. Obtained mixture was taken for sample after silver powder precipitation.

Sample was placed between two KBr glass plates with identical clamps. In the capacity of comparative sample were used empty glass plates. Every sample spectrum was performed for five times and the final one was calculated by striking an average. Procedure was carried out in transmission mode and then data were recomputed to obtain absorption spectra.

There are four possible states of AA: ketone, ketone-enol, enol A, and enol B (Figure 5.12). Vibrational analysis is carried out with one of semi-empirical of *ab initio* methods and can allow recognizing different states of reagents by spectra comparison.

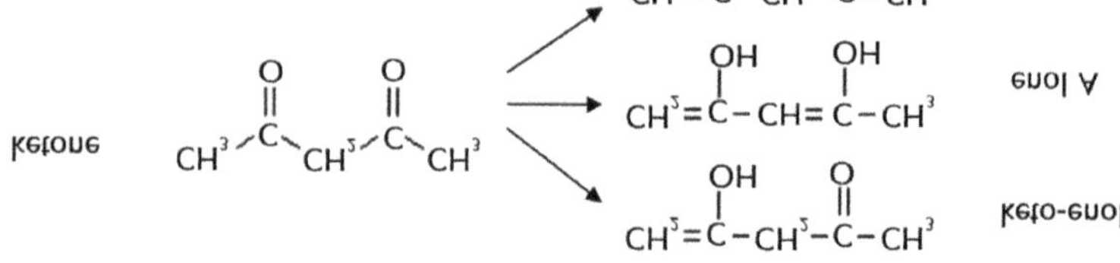

FIGURE 5.12   Different states of acetylacetone.

The vibrational frequencies are derived from the harmonic approximation, which assumes that the potential surface has a quadratic form. The association between transition energy ΔE and frequency v is performed by Einstein's formula

$$\Delta E = h \cdot v \quad (5.2)$$

where, h is the Planck constant. IR frequencies ($\sim 10^{12}$Hz) accord with gaps between vibrational energy levels. Thus, each line in an IR spectrum represents an excitation of nuclei from one vibrational state to another [17].

Comparison between MNDO, AM1 and PM3 methods were performed in [18]. Semi-empirical method PM3 demonstrated the closest correspondence to experimental values.

In the Figure 5.13 showed experimental data of IR-Fourier spectrometer AA measurement and data achieved from the National Institute of Advanced Industrial Science and Technology open database. Peaks comparison is clearly shown both spectra generally identical. IR spectra for comparison with calculated ones were received from open AIST database [1]. As it can be concluded from AIST data, experimental IR spectrum was measured for AA ketone state.

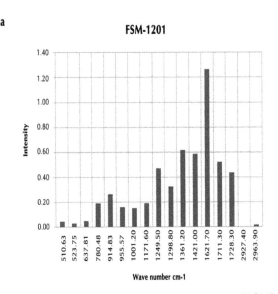

FIGURE 5.13 (*Continued*)

b

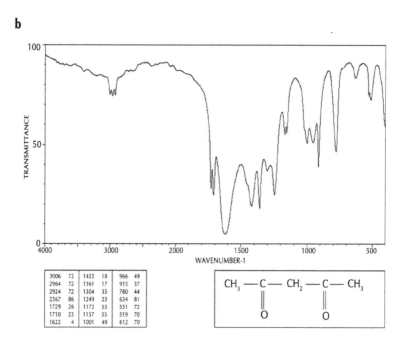

3006	72	1422	18	966	49
2964	72	1361	17	915	37
2924	72	1304	35	780	44
2367	86	1249	23	634	81
1729	26	1172	53	531	72
1710	23	1157	55	519	70
1622	4	1001	49	612	70

FIGURE 5.13 Experimental IR spectrum of acetylacetone obtained from IR-Fourier FSM-1201 spectrometer (a) and IR spectrum of acetylacetone received from open AIST database (b).

As one can see, most intensive peak 1,622cm^{-1} belongs to C=O bond vibrations. Oxygen relating to this bond can participate in different coordination reactions. Peaks 915cm^{-1} and 956cm^{-1} belong to O–H bond vibrations. This bond can also take part in many reactions.

Computational part is comprised several steps. There were several steps in computational part. Firstly, geometry of AA was optimized to achieve molecule stable state. Then vibrational analysis was carried out. Parameters of the same computational method were used for subsequent IR spectrum calculation. Obtained IR spectra are shown in Figure 5.14.

a)

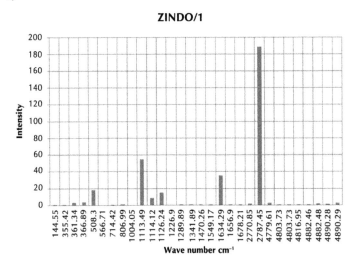

b)

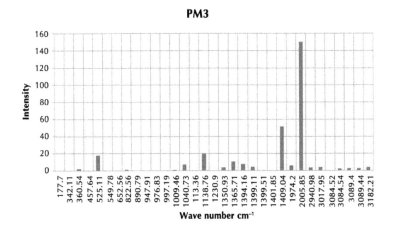

FIGURE 5.14   (Continued)

c)

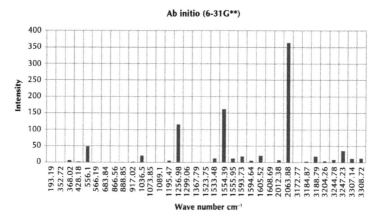

d)

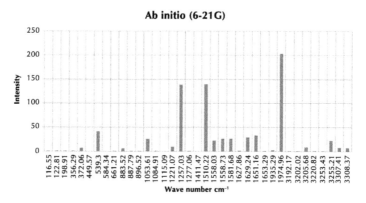

FIGURE 5.14   IR spectra of acetylacetone calculated with HyperChem software: semi-empirical method ZINDO/1 (a); semi-empirical method PM3 (b); *ab initio* method with 6-31G** basis set (c); *ab initio* method with 6-21G basis set (d)

Vibrational analysis performed by HyperChem was showed that PM3 demonstrates more close correspondence with experimental and *ab initio* calculated spectra than ZINDO/1. As expected, *ab initio* spectra were demonstrated closest result in general. As we can see at Figure 5.14c, d, peaks of IR spectrum calculated with 6-21G small basis set are staying closer with respect to experimental values than ones calculated with large basis set 6-31G**. Thus it will be reasonable to use both PM3 and *ab initio* with 6-21G basis set methods in further calculations.

a)

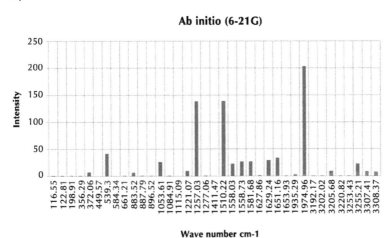

b)

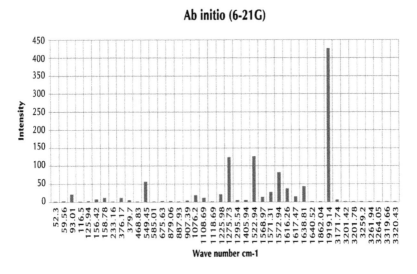

FIGURE 5.15   (*Continued*)

c)

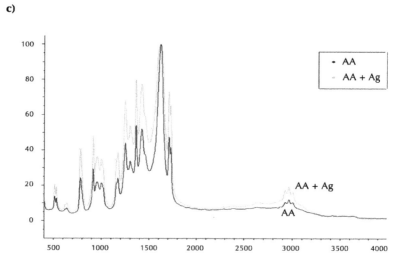

FIGURE 5.15 IR spectra obtained with *ab initio* method with 6-21G basis set: acetylacetone (a); acetylacetone with silver powder (b); experimental data obtained from IR-Fourier FSM-1201 spectrometer: acetylacetone with silver powder in comparison with pure acetylacetone (c)

As shown on Figure 5.15a, b, peaks are situated almost identical due to each other with the exception of peak 1,573cm^{-1} (Figure 5.15b) which was appeared after silver powder addition. Peak 1,919cm^{-1} (Figure 5.15b) has been more intensive as compared with similar one on Figure 5.15a.

Peaks are found in range 600–700cm^{-1} should usually relate to metal complexes with acetylacetone [19, 20]. IR spectra AA and AA with Ag^{+} ion calculations were carried out by *ab initio* method with 6-21G basis set. PM3 wasn't used due this method hasn't necessary parametrization for silver. In the case of IR spectra *ab initio* computation unavoidable calculating error is occurred hence all peaks have some displacement. Peak 549cm^{-1} (Figure 5.15b) is equivalent to peak 638cm^{-1} (Figure 5.15c) obtained experimentally and it is more intensive than similar one on Figure 5.4. Peak 1,974cm^{-1} was displaced to mark 1,919cm^{-1} and became far intensive. It can be explained Ag^{+} influence and coordination bonds between metal and AA formation.

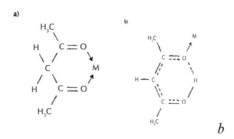

FIGURE 5.16 Different ways of complex between metal and AA formation: ketone with metal (a); enol with metal (b)

Complexes between silver and AA depending on enol or ketone state can form with two different ways (Figure 5.16).

IR spectra of pure acetylacetone and acetylacetone with silver were calculated with different methods of computational chemistry. PM3 method was shown closest results with respect to experimental data among semi-empirical methods. It was established that there is no strong dependence between size of *ab initio* method basis set and data accuracy. It was found that IR spectrum carried out with *ab initio* method with small basis set 6-21G has more accurate data than analogous one with large basis set 6-31G**. Experimental and calculated spectra were shown identical picture of spectral changes with except of unavoidable calculating error. Hence, PM3 and *ab initio* methods can be used in metal–carbon complexes formations investigations.

5.2   THE METAL OR CARBON NANOCOMPOSITES INFLUENCE MECHANISMS ON MEDIA AND ON COMPOSITIONS

The modification of materials by nanostructures including metal / carbon nanocomposites consists in the conducting of the following stages:

1. The choice of liquid phase for the making of finely dispersed suspensions intended for the definite material or composition.
2. The making of finely dispersed suspension with sufficient stability for the definite composition.
3. The development of conditions of the finely dispersed suspension introduction into composition.

At the choice of liquid phase for the making of finely dispersed suspension it should be taken the properties of nanostructures (nanocomposites) as well as liquid phase (polarity, dielectric penetration, viscosity).

It is perspective if the liquid phase completely enters into the structure of material formed during the composition hardening process.

When the correspondent solvent, on the base of which the suspension is obtained, is evaporated, the re-coordination of nanocomposite on other components takes place and the effectiveness of action of nanostructures on composition is decreased.

The stability of finely dispersed suspension is determined on the optical density. The time of the suspension optical density conservation defines the stability of suspension. The activity of suspension is found on the bands intensity changes by means of IR and Raman spectra. The intensity increasing testifies to transfer of nanostructure surface energy vibration part on the molecules of medium or composition. The line speeding in spectra testify to the growth of electron action of nanocomposites with medium molecules. Last fact is confirmed by x-ray photoelectron investigations.

The changes of character of distribution on nanoparticles sizes take place depending on the nature of nanocomposites, dielectric penetration and polarity of liquid phase.

Below characteristics of finely dispersed suspensions of metal or carbon nanocomposites are given.

The distribution of nanoparticles in water, alcohol and water-alcohol suspensions prepared based on the above technique are determined with the help of laser analyzer. In Figures 5.17 and 5.18 one can see distributions of copper or carbon nanocomposite in the media different polarity and dielectric penetration.

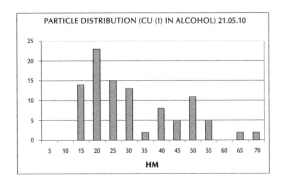

FIGURE 5.17 Distribution of copper /carbon nanocomposites in alcohol

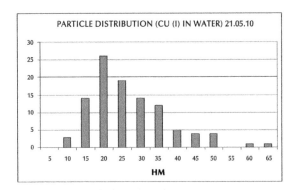

FIGURE 5.18 Distribution of copper / carbon nanocomposites in water

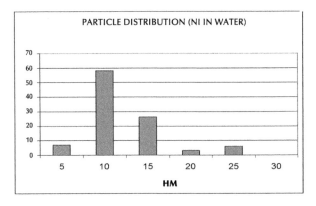

FIGURE 5.19 Distribution of nickel or carbon nanocomposites in water

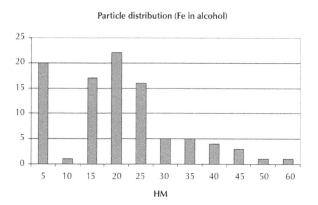

FIGURE 5.20   Distribution of iron/ carbon nanocomposites

When comparing the figures we can see that ultrasound dispergation of one and the same nanocomposite in media different by polarity results in the changes of distribution of its particles. In water solution the average size of Cu/C nanocomposite equals 20nm, and in alcohol medium—greater than 5nm.

Assuming that the nanocomposites obtained can be considered as oscillators transferring their oscillations onto the medium molecules, we can determine to what extent the IR spectrum of liquid medium will change, for example polyethylene polyamine applied as a hardener in some polymeric compositions, when we introduce small and supersmall quantities of nanocomposite into it.

IR spectra demonstrate the change in the intensity at the introduction of metal or carbon nanocomposite in comparison with the pure medium (IR spectra are given in Figure 5.21). The intensities of IR absorption bands are directly connected with the polarization of chemical bonds at the change in their length, valence angles at the deformation oscillations, that is, at the change in molecule normal coordinates.

When nanostructures are introduced into media, we observe the changes in the area and intensity of bands, that indicates the coordination interactions and influence of nanostructures onto the medium (Figures 5.21–5.23).

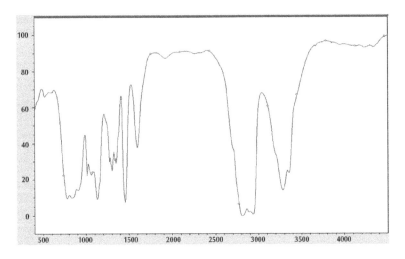

FIGURE 5.21   IR spectrum of polyethylene polyamine

Special attention in PEPA spectrum should be paid to the peak at 1,598cm^{-1} attributed to deformation oscillations of N–H bond, where hydrogen can participate in different coordination and exchange reactions.

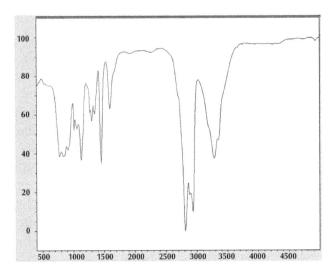

FIGURE 5.22   IR spectrum of copper or carbon nanocomposite finely dispersed suspension in polyethylene polyamine medium (ω (NC) = 1%)

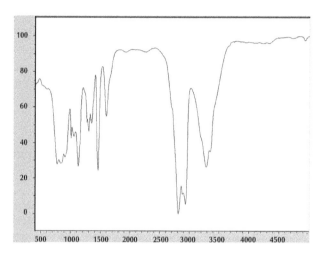

FIGURE 5.23 IR spectrum of nickel or carbon nanocomposite fine suspension in polyethylene polyamine (ω (NC) = 1%)

In the spectra wave numbers characteristic for symmetric $v_s(NH_2)$ 3,352cm^{-1} and asymmetric $v_{as}(NH_2)$ 3,280cm^{-1} oscillations of amine groups are present. There is a number of wave numbers attributed to symmetric $v_s(CH_2)$ 2,933cm^{-1} and asymmetric valence $v_{as}(CH_2)$ 2,803cm^{-1}, deformation wagging oscillations $vD(CH_2)$ 1,349cm^{-1} of methylene groups, deformation oscillations of NH vD (NH) 1,596cm^{-1} and NH_2 $vD(NH_2)$ 1,456cm^{-1} amine groups. The oscillations of skeleton bonds at $v(CN)$ 1,059–1,273cm^{-1} and $v(CC)$ 837cm^{-1} are the most vivid. The analysis of intensities of IR spectra of PEPA and fine suspensions of metal or carbon nanocomposites based on it revealed a significant change in the intensities of amine groups of dispersion medium (for $v_s(NH_2)$ in 1.26 times, and for $v_{as}(NH_2)$ in approximately 50 times).

Such demonstrations are presumably connected with the distribution of the influence of nanoparticle oscillations onto the medium with further structuring and stabilizing of the system. Under the influence of nanoparticle the medium changes which is confirmed by the results of IR spectroscopy (change in the intensity of absorption bands in IR region). Density, dielectric penetration, and viscosity of the medium are the determining parameters for obtaining fine suspension with uniform distribution of particles in the volume. At the same time, the structuring rate and consequently the stabilization of the system directly depend on the distribution by particle sizes in suspension. At the wide range of particle distribution by sizes, the oscillation frequency of particles different by size can significantly differ, in this connection; the distortion in the influence transfer of nanoparticle system onto the medium is possible (change in the medium from the part of some particles can be balanced by the other). At the narrow range of nanoparticle distribution by sizes the system structuring and stabilization are possible. With further adjustment of the components such processes will positively influence the processes of structuring and self-organization of final composite system determining physical-mechanical characteristics of hardened or hard composite system.

The effects of the influence of nanostructures at their interaction into liquid medium depend on the type of nanostructures, their content in the medium and medium nature. Depending on the material modified, fine suspensions of nanostructures based on different media are used. Water and water solutions of surface-active substances, plasticizers, and foaming agents (when modifying foam concretes) are applied as such media to modify silicate, gypsum, cement, and concrete compositions. To modify epoxy compounds and glues based on epoxy resins the media based on polyethylene polyamine, isomethyltetrahydrophthalic anhydride, toluene, and alcohol-acetone solutions are applied. To modify polycarbonates and derivatives of polymethyl methacrylate dichloroethane and dichloromethane media are used. To modify polyvinyl chloride compositions and compositions based on phenolformaldehyde and phenolrubber polymers alcohol or acetone-based media are applied. Fine suspensions of metal or carbon nanocomposites are produced using the above media for specific compositions. In IR spectra of all studied suspensions the significant change in the absorption intensity, especially in the regions of wave numbers close to the corresponding nanocomposite oscillations, is observed. At the same time, it is found that the effects of nanocomposite influence on liquid media (fine suspensions) decreases with time and the activity of the corresponding suspensions drops. The time period in which the appropriate activity of nanocomposites is kept changes in the interval 24 hrs–1 month depending on the nanocomposite type and nature of the basic medium (liquid phase in which nanocomposites dispergate). For instance, IR spectroscopic investigation of fine suspension based on isomethyltetrahydrophthalic anhydride containing 0.001% of Cu/C nanocomposite indicates the decrease in the

peak intensity, which sharply increased on the third day when nanocomposite
was introduced (Figure 5.24).

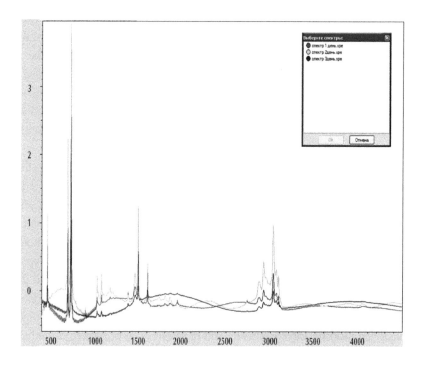

FIGURE 5.24 Changes in IR spectrum of copper or carbon nanocomposite fine
suspension based on isomethyltetrahydrophthalic anhydride with time. (a) IR spectrum on
the first day after the nanocomposite was introduced, (b)IR spectrum on the second day,
(c) IR spectrum on the third day

Similar changes in IR spectra take place in water suspensions of metal or carbon nanocomposites based on water solutions of surface-active
nanocomposites.

In Figure 5.25 one can see IR spectrum of iron or carbon nanocomposite based on water solution of sodium lignosulfonate in comparison with
IR spectrum of water solution of surface-active substance.

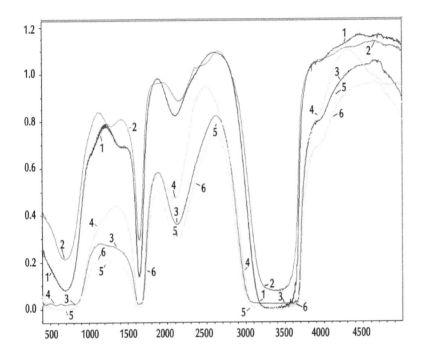

FIGURE 5.25 Comparison of IR spectra of water solution of sodium lignosulfonate (*1*) and fine suspension of iron or carbon nanocomposite (0.001%) based on this solution on the first day after nanocomposite introduction (*2*), on the third day (*3*), on the seventh day (*4*), 14th day (*5*) and 28th day (*6*)

As it is seen, when nanocomposite is introduced and undergoes ultra-sound dispergation, the band intensity in the spectrum increases signifi-cantly. Also the shift of the bands in the regions 1,100–1,300cm^{-1}, 2,100–2,200cm^{-1} is observed, which can indicate the interaction between sodium lignosulfonate and nanocomposite. However after two weeks the decrease in band intensity is seen. As the suspension stability evaluated by the optic density is 30 days, the nanocomposite activity is quite high in the period when IR spectra are taken. It can be expected that the effect of foam con-crete modification with such suspension will be revealed if only 0.001% of nanocomposite is introduced.

5.3   THE COMPOSITIONS MODIFICATION PROCESSES BY METAL / CARBON NANOCOMPOSITES

5.3.1   GENERAL FUNDAMENTALS OF POLYMERIC MATERIALS MODIFICATION BY METAL / CARBON NANOCOMPOSITES

The material modification with the using of metal or carbon nanocomposites is usually carried out by finely dispersed suspensions containing solvents or components of polymeric compositions. We realize the modification of the following materials: concrete foam, dense concrete, water glass, polyvinyl acetate, polyvinyl alcohol, polyvinyl chloride, polymethyl methacrylate, polycarbonate, epoxy resins, phenol-formaldehyde resins, reinforced plastics, glues, pastes, including current conducting polymeric materials and filled polymeric materials. Therefore, it is necessary to use the different finely dispersed suspension for the modification of enumerated materials. The series of suspensions consist of the suspensions on the basis of following liquids: water, ethanol, acetone, benzene, toluene, dichlorethane, methylene chloride, oleic acid, polyethylene polyamine, isomethyl tetra hydrophtalic anhydrite, water solutions surface-active substances, or plasticizers. In some cases, the solutions of correspondent polymers are applied for the making of the stable finely dispersed suspensions. The estimation of suspensions stability is given as the change of optical density during the definite time (Figure 5.26)

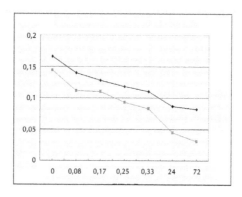

FIGURE 5.26   The change of optical density of typical suspension depending on time

Relative change of free energy of coagulation process also may be as the estimation of suspension stability

$$\Delta\Delta F = lg\ k_{NC}/k_{NT} \qquad (5.3)$$

where, k_{NC}—exp(-ΔF_{NC}/RT)—the constant of coagulation rate of nano-composite, k_{NT}—exp(-ΔF_{NT}/RT)—the constant of coagulation rate of carbon nanotube, ΔF_{NC}, ΔF_{NT}—the changes of free energies of corresponding systems (Figure 5.27).

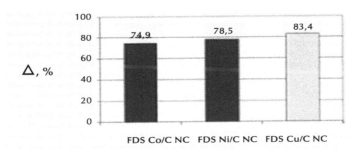

FDS Co/C NC   FDS Ni/C NC   FDS Cu/C NC

FIGURE 5.27   The comparison of finely dispersed suspensions stability

The stability of metal or carbon nanocomposites suspension depends on the interactions of solvents with nanocomposites participation (Figure 5.28).

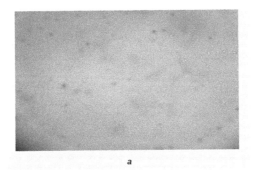

a

FIGURE 5.28   (*Continued*)

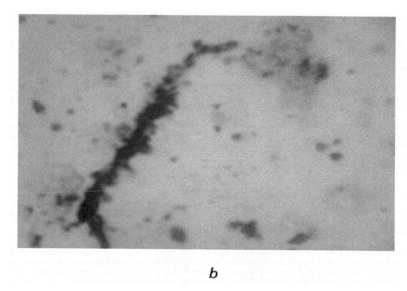

b

FIGURE 5.28 Microphotographs of Co/C nanocomposite suspension on the basis of mixture "dichlorethane–oleic acid" (a) and oleic acid (b)

The introduction of metal or carbon nanocomposites leads to the changes of kinematic and dynamic viscosity (Figure 5.29).

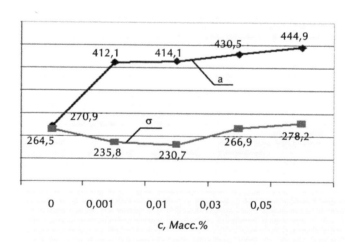

FIGURE 5.29 The dependence of dynamic and kinematic viscosity on the Cu/C nanocomposite quantity

For the description of polymeric composition self organization the critical parameters (critical content of nanostructures critical time of process realization, critical energetic action) may be used.

The equation of nanostructure influence on medium is proposed

$$W = n/N \; \exp\{an\tau^{\beta}/T\} \tag{5.4}$$

where, n—number of active nanostructures, N—number of nanostructure interaction, a—activity of nanostructure, τ—duration of self organization process, T—temperature, —degree of freedom (number of process direction).

5.3.2 ABOUT THE MODIFICATION OF FOAM CONCRETE BY METAL OR CARBON NANOCOMPOSITES SUSPENSION

The ultimate breaking stresses were compared in the process of compression of foam concretes modified with copper / carbon nanocomposites obtained in different nanoreactors of polyvinyl alcohol [21, 22]. The sizes of nanoreactors change depending on the crystallinity and correlation of acetate and hydroxyl groups in PVA which results in the change of sizes and activity of nanocomposites obtained in nanoreactors. It is observed that the sizes of nanocomposites obtained in nanoreactors of PVA matrixes 16/1(ros) (NC2), PVA 16/1 (imp) (NC1), PVA 98/10 (NC3), correlate as NC3 > NC2 > NC1. The smaller the nanoparticle size and greater its activity, the fewer amounts of nanostructures is required for self-organization effect.

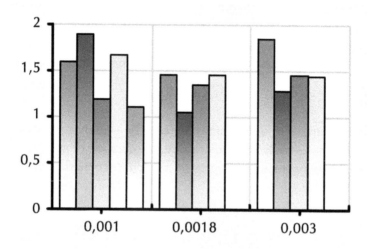

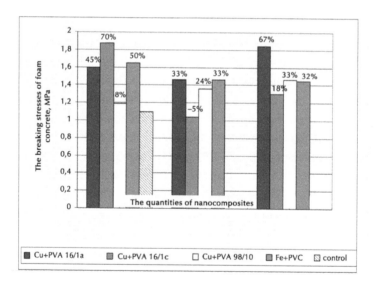

FIGURE 5.30 The dependence of breaking stresses on quantity (%) of metal or carbon nanocomposites

At the same time, the oscillatory nature of the influence of these nano-composites on the compositions of foam concretes is seen in the fact that if the amount of nanocomposite is 0.0018% from the cement mass, the significant decrease in the strength of NC1 and NC2 is observed. The increase in foam concrete strength after the modification with iron or carbon nanocomposite is a little smaller in comparison with the effects after the application of NC1 and NC2 as modifiers. The corresponding effects after the modification of cement, silicate, gypsum and concrete compositions with nanostructures is defined by the features of components and technologies applied. These features often explain the instability of the results after the modification of the foregoing compositions with nanostructures. Besides, during the modification the changes in the activity of fine suspensions of nanostructures depending on the duration and storage conditions should be taken into account.

In this regard, it is advisable to use metal / carbon nanocomposites when modifying polymeric materials whose technology was checked on strictly controlled components.

In the paper the possibilities of developing new ideas about self-organization processes and about nanostructures and nanosystems are discussed on the example of metal or carbon nanocomposites. The obtaining of metal / carbon nanocomposites in nanoreactors of polymeric matrixes is proposed to consider as self-organization process similar to the formation of ordered phases which can be described with Avrami's equation. The application of Avrami's equations during the synthesis of nanofilm structures containing copper clusters has been tested. The influence of nanostructures on active media is given as the transfer of oscillation energy of the corresponding nanostructures onto the medium molecules.

IR spectra of metal/ carbon and their finely dispersed suspensions in different media (water and organic substances) have been studied for the first time. It has been found that the introduction of super small quantities of prepared nanocomposites leads to the significant change in band intensity in IR spectra of the media. The attenuation of oscillations generated by the introduction of nanocomposites after the time interval specific for the pair "nanocomposite–medium" has been registered.

Thus to modify compositions with finely dispersed suspensions it is necessary for the latter to be active enough that should be controlled with IR spectroscopy.

A number of results of material modification with finely dispersed suspensions of metal /carbon nanocomposites are given, as well as the examples of changes in the properties of modified materials based on concrete compositions, epoxy and phenol resins, polyvinyl chloride, polycarbonate, and current-conducting polymeric materials.

5.3.3   THE MODIFICATION OF EPOXY RESINS BY METAL OR CARBON NANOCOMPOSITES SUSPENSIONS

IR spectra of polyethylene polyamine and metal or carbon nanocomposite fine suspensions based on it with concentrations 0.001–0.03% from polyethylene polyamine mass are analyzed and their comparative analysis is presented. The possible processes flowing in fine suspensions during the interaction of copper/ carbon nanocomposite and polyethylene polyamine are described, as well as the processes influencing the increase in adhesive strength and thermal stability of metal / carbon nanocomposite / epoxy compositions.

This decade has been heralded by a large-scale replacement of conventional metal structures with structures from polymeric composite materials (PCM). Currently we are facing the tendency of production growth of PCM with improved operational characteristics. In practice, PCM characteristics can be improved when applying modern manufacturing technologies, for example, the application of "binary" technologies of prepreg production [23], as well as the synthesis of new polymeric PCM matrixes or modification of the existing polymeric matrixes with different fillers.

The most cost-efficient way to improve operational characteristics is to modify the existing polymeric matrixes; therefore, currently the group of polymeric materials modified with nanostructures (NS) is of special interest. NS are able to influence the supermolecular structure, stimulate self-organization processes in polymeric matrixes in supersmall quantities, thus contributing to the efficient formation of a new phase "medium modified—nanocomposite" and qualitative improvement of the characteristics of final product—PCM. This effect is especially visible when NS activity increases which directly depends on the size of specific surface, shape of the particle and its ultimate composition [22]. Metal ions in the NS used in this work also contribute to the activity increase as they stimulate the formation of new bonds.

The increase in the attraction force of two particles is directly proportional to the growth of their elongated surfaces, therefore the possibility of NS coagulation and decrease in their efficiency as modifiers increases together with their activity growth. This fact and the fact that the effective concentrations to modify polymeric matrixes are usually in the range below 0.01 mass % impose specific methods for introducing NS into the material modified. The most justified and widely applicable are the methods for introducing NS with the help of fine suspensions (FS).

Quantum-chemical modeling methods allow quite precisely defining the typical reaction of interaction between the system components, predict the properties of molecular systems, decrease the number of experiments due to the imitation of technological processes. The computational results are completely comparable with the experimental modeling results.

The optimal composition of nanocomposite was found with the help of software HyperChem v. 6.03. Cu/C nanocomposite is the most effective for modifying the epoxy resin. Nanosystems formed with this NC have higher interaction energy in comparison with nanosystems produced with Ni/C and Co/C nanocomposites. The effective charges and geometries of nanosystems were found with semi-empirical methods. The fact of producing stable fine suspensions of nanocomposite on PEPA basis and increasing the operational characteristics of epoxy polymer was ascertained. It was demonstrated that the introduction of Cu/C nanocomposite into PEPA facilitates the formation of coordination bonds with nitrogen of amine groups, thus resulting in PEPA activity increase in epoxy resin hardening reactions.

It was found out that the optimal time period of ultrasound processing of copper / carbon nanocomposite fine suspension is 20 mins.

The dependence of Cu/C nanocomposite influence on PEPA viscosity in the concentration range 0.001–0.03% was found. The growth of specific surface of NC particles contributes to partial decrease in PEPA kinematic viscosity at concentrations 0.001, 0.01% with its further elevation at the concentration 0.03%.

IR investigation of Cu/C nanocomposite fine suspension confirms the quantum-chemical computational experiment regarding the availability of NC interactions with PEPA amine groups. The intensity of these groups increased in several times when Cu/C nanocomposite was introduced.

The test for defining the adhesive strength and thermal stability correlate with the data of quantum-chemical calculations and indicate the formation of a new phase facilitating the growth of cross-links number in polymer grid when the concentration of Cu/C nanocomposite goes up. The optimal concentration for elevating the modified epoxy resin (ER) adhesion equals 0.003% from ER weight. At this concentration the strength growth is 26.8%. From the concentration range studied, the concentration 0.05% from ER weight is optimal to reach a high thermal stability. At this concentration the temperature of thermal destruction beginning increases up to 195°C.

Thus, in this work the stable fine suspensions of Cu/C nanocomposite were obtained. The modified polymers with increased adhesive strength (by 26.8%) and thermal stability (by 110°C) were produced based on epoxy resins and fine suspensions.

5.3.4 THE MODIFICATION OF GLUES BASED ON PHENOL-FORMALDEHYDE RESINS BY METAL OR CARBON NANOCOMPOSITES FINELY DISPERSED SUSPENSION

In modern civil and industrial engineering, mechanical engineering and extra strong, rather light, durable metal structures, wooden structures, and light assembly structures are widely used. However, they deform, lose stability and load-carrying capacity under the action of high temperatures, therefore, fire protection of such structures is important issue for investigation.

To decrease the flammability of polymeric coatings it is advisable to create and forecast the properties of materials with external intumescent coating containing active structure-forming agents—regulators of foam cokes structure. At present, metal/ carbon nanocomposites are such perspective modifiers. The introduction of nanocomposites into the coating can improve the material behavior during the combustion, retarding the combustion process [24, 25].

In this regard, it is advisable to modify metal / carbon nanocomposites with ammonium phosphates to increase the compatibility with the corresponding compositions. The grafting of additional functional groups is appropriate in strengthening the additive interaction with the matrix and, thus, in improving the material properties. The grafting can contribute to improving the homogeneity of nanostructure distribution in the matrix and suspension stability. Due to partial ionization additional groups produce a small surface charge with the result of nanostructure repulsion from each other and stabilization of their dispersion. Phosphorylation leads to the increase in the influence of nanostructures on the medium and material being modified, as well as to the improvement of the quality of nanostructures due to metal reduction [26–28].

The glue BF-19 (based on phenol-formaldehyde resins) is intended for gluing metals, ceramics, glass, wood and fabric in hot condition, as well as for assembly gluing of cardboard, plastics, leather and fabrics in cold condition. The glue composition: organic solvent, synthetic resin (phenol-formaldehyde resins of new lacquer type), synthetic rubber.

When modifying the glue composition, at the first stage the mixture of alcohol suspension (ethyl alcohol + Me/C NC modified) and ammonium polyphosphate (APPh) was prepared. At the same time, the mixtures containing ethyl alcohol, Me/C NC and APPh, ethyl alcohol and APPh were prepared. At the second stage the glue composition was modified by the introduction of phosphorus containing compositions prepared into the glue BF-19.

The interaction of nanocomposites with ammonium polyphosphate results in grating phosphoryl groups to them that allowed using these nanocomposites to modify intumescent fireproof coatings.

The sorption ability of functionalized nanocomposites was studied. It was found out that phospholyrated nanostructures have higher sorption ability than nanocomposites not containing phosphorus. Since the sorption ability indicates the nanoproduct activity degree, it can be concluded that nanocomposite activity increases in the presence of active medium and modifier (ammonium polyphosphate).

IR spectra of suspensions of nanocomposites on alcohol basis were investigated. It was found that changes in IR spectrum by absorption intensity indicate the presence of excitation source in the suspension. It was assumed that nanostructures are such source of band intensity increase in suspension IR spectra.

The glue coatings were modified with metal / carbon nanocomposite. It was determined that nanocomposite introduction into the glue significantly decreases the material flammability. The samples with phospholyrated nanocomposites have better test results. When phosphorus containing nanocomposite is introduced into the glue, foam coke is formed on the sample surface during the fire exposure. The coating flaking off after flame exposure was not observed as the coating preserved good adhesive properties even after the flammability test.

The nanocomposite surface phospholyration allows improving the nanocomposite structure, increases their activity in different liquid media thus increasing their influence on the material modified. The modification of coatings with nanocomposites obtained finally results in improving their fire-resistance and physical and chemical characteristics.

5.3.5 THE MODIFICATION OF POLYCARBONATE BY METAL/CARBON NANOCOMPOSITES

Recently the materials based on polycarbonate have been modified to improve their thermal-physical and optical characteristics and to apply new properties to use them for special purposes. The introduction of nanostructures into materials facilitates self-organizing processes in them. These processes depend on surface energy of nanostructures which is connected with energy of their interaction with the surroundings. It is known [21] that the surface energy and activity of nanoparticles increase when their sizes decrease.

For nanoparticles the surface and volume are defined by the defectiveness and form of conformation changes of film nanostructures depending on their crystallinity degree. However, the possibilities of changes in nanofilm shapes with the changes in medium activity are greater in comparison with nanostructures already formed. At the same time, sizes of nanofilms formed and their defectiveness, that is, tears and cracks on the surface of nanofilms, play an important role [22]

When studying the influence of supersmall quantities of substances introduced into polymers and considerably changing their properties, apparently we should consider the role which these substances play in polymers possessing highly organized supermolecular regularity both in crystalline and amorphous states. It can be assumed that the mechanism of this phenomenon is in the nanostructure energy transfer to polymer structural formations through the interface resulting in the changes in their surface energy and mobility of structural elements of the polymeric body. Such mechanism is quite realistic as polymers are structural-heterogenic (highly dispersed) systems.

Polycarbonate modification with supersmall quantities of Cu/C nanocomposite is possible using fine suspensions of this nanocomposite which contributes to uniform distribution of nanoparticles in polycarbonate solution.

Polycarbonate "Actual" was used as the modified polycarbonate.

Fine suspensions of copper / carbon nanocomposite were prepared combining 1.0, 0.1, 0.01, and 0.001% of nanocomposite in polycarbonate solution in ethylene dichloride. The suspensions underwent ultrasonic processing.

To compare the optical density of nanocomposite suspension in polycarbonate solution in ethylene dichloride, as well as polycarbonate and polycarbonate samples modified with nanocomposites, spectrophotometer KFK-3-01a was used.

Samples in the form of modified and non-modified films for studying IR spectra were prepared precipitating them from suspension or solution under vacuum. The obtained films about 100mcm thick were examined on Fourier-spectrometer FSM 1201.

To investigate the crystallization and structures formed the high-resolution microscope (up to 10mcm) was used.

To examine thermal-physical characteristics the lamellar material on polycarbonate and polymethyl methacrylate basis about 10mm high was prepared. Three layers of polymethyl methacrylate and two layers of polycarbonate were used. Thermal-physical characteristics (specific thermal capacity and thermal conductivity) were investigated on calorimeters IT-c-400 and IT-λ-400.

During the investigation the nanocomposite fine suspension in polycarbonate solution in ethylene dichloride was studied, polycarbonate films modified with different concentrations of nanocomposite were compared with the help of optical spectroscopy, microscopy, IR spectroscopy and thermal-physical methods of investigation.

The results of investigation of optical density of Cu/C nanocomposite fine suspension (0.001%) based on polycarbonate solution in ethylene dichloride are given in Figure 5.31. As seen in the figure, the introduction of nanocomposite and polycarbonate into ethylene dichloride resulted in transmission increase in the range 640–690nm (approximately in three times). At the same time, at 790 and 890nm the significant increase in optical density was observed. The comparison of optical densities of films of polycarbonate and modified materials after the introduction of different quantities of nanocomposite (1.0, 0.1, and 0.01%) into polycarbonate is interesting (Figure 5.32).

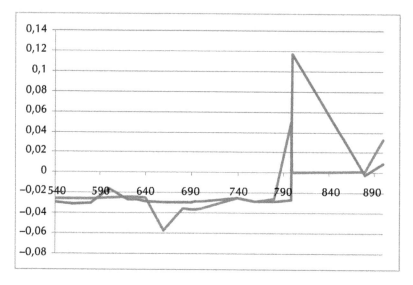

FIGURE 5.31 Curves of optical densit y of suspension based on ethylene dichloride diluted with polycarbonate and Cu/C nanocomposite in the concentration to polycarbonate 0.001% (1) and ethylene dichloride (2)

Comparison of optical density of suspension containing 0.001% of nanocomposite and optical density polycarbonate sample modified with 0.01% of nanocomposite indicates the proximity of curves character. Thus the correlation of optical properties of suspensions of nanocomposites and film materials modified with the same nanocomposites is quite possible.

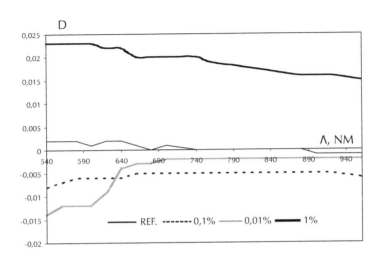

FIGURE 5.32   Curves of optical density of reference sample modified with Cu/C nanocomposites in concentration 1, 0.1, and 0.01%

Examination of curves of samples optical density demonstrated that when the nanostructure concentration was 1% from polycarbonate mass, the visible-light spectrum was absorbed by about 4.2% more if compared with the reference sample. When the nanocomposite concentration was 0.1% the absorption decreased by 0.7%. When the concentration was 0.01%, the absorption decreased in the region 540–600nm by 2.3%, and in the region 640–960nm—by 0.5%.

During the microscopic investigation of the samples the schematic picture of the structures formed was obtained at 20-mcm magnification. The results are given in Figure 5.33. From the schematic pictures it is seen that volumetric structures of regular shape surrounded by micellae were formed in polycarbonate modified with 0.01% Cu/C nanocomposite. When 0.1% of nanocomposite was introduced, the linear structures distorted in space and surrounded by micellae were formed. When Cu/C nanocomposite concentration was 1%, large aggregates were not observed in polycarbonate.

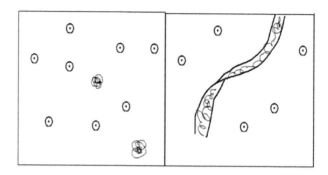

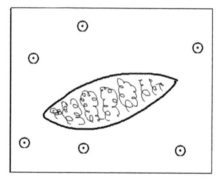

FIGURE 5.33   Schematic picture of structure formation in polycarbonate modified with Cu/C nanocomposites in concentrations 1, 0.1, and 0.01% (from *left* to the *right*), 20-mcm magnification

Apparently, the decrease in nanocomposite concentration in polycarbonate can result in the formation of self-organizing structures of bigger size. In [3] there is the hypothesis on the transfer of nanocomposite oscillations onto the molecules of polymeric composition, the intensity of bands in IR spectra of which sharply increases even after the introduction of supersmall quantities of nanocomposites. In our case, this hypothesis was checked on modified and non-modified samples of polycarbonate films.

In Figure 5.34 one can see the IR spectra of polycarbonate and polycarbonate modified with 0.001% of Cu/C nanocomposite.

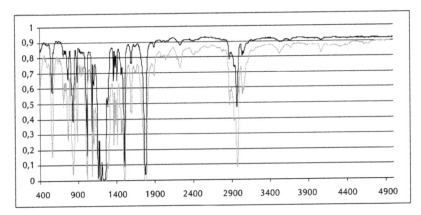

FIGURE 5.34 IR spectra of reference sample (*upper*) and polycarbonate modified with Cu/C nanocomposites in concentration 0.001% (*lower*)

As seen in IR spectra, the intensity increases in practically all regions, indicating the influence of oscillations of Cu/C nanocomposite on all the system. The most vivid changes in the band intensity are observed at $557cm^{-1}$, in the region $760–930cm^{-1}$, at $1,400cm^{-1}$, $1,600cm^{-1}$, $2,970cm^{-1}$.

Thus, polycarbonate self-organization under the influence of Cu/C nanocomposite takes place with the participation of certain bonds, for which the intensity of absorption bands goes up. At the same time, the formation of new phases is possible which usually results in thermal capacity increase. But thermal conductivity can decrease due to the formation of defective regions between the aggregates formed.

The table contains the results of thermal-physical characteristics studied.

TABLE 5.11 Thermal-physical characteristics of polycarbonate and its modified analogs

Parameters	Nanocomposite content in polycarbonate (%)			
	0 (ref.)	1	0.1	0.01
m 10-3, kg	1.9	1.955	1.982	1.913
h 10-3, m	9.285	9.563	9.76	9.432
Csp, J/kg K	1440	1028	1400	1510
λ	0.517	0.503	0.487	0.448

It is demonstrated that when the nanostructure concentration in the material decreases, thermal capacity goes up that is confirmed by the results of previous investigations. Thermal conductivity decline, when the nanostructure concentration decreases, is apparently caused by the material defectiveness.

When Cu/C nanocomposites are introduced into the material modified, the nanostructures can be considered as the generator of molecules excitation, which results in wave process in the material.

It is found that polycarbonate modification with metal or carbon containing nanocomposites results in the changes in polycarbonate structure influencing its optical and thermal-physical properties.

KEYWORDS

- α-Naphthol molecules
- Heptagonal defect
- Hydroxyl fullerene
- Interaction energy
- Quantum-chemical investigation

REFERENCES

1. Schmidt, M. W., Baldridge, K. K., & Boatz, J. A. et al. (1993). *Journal of Computational Chemistry, 14,* 1347–1363.
2. Copley, J. R. D., Neumann, D. A., Cappelletti, R. L., & Kamitakahara, W. A. (1992). *Journal of Physics and Chemistry of Solids, 53*(11), 1353–1371.
3. Lo, Sh., & Li, V. (1999). *Russian Chemical Journal, 5,* 40–48.
4. Ponomarev, A. N. (2007). *Constructional Materials, 6,* 69–71.
5. Krutikov, V. A., Didik, A. A., Yakovlev, G. I., Kodolov, V. I., & Bondar, A. Yu. (2005). *Alternative Power Engineering and Ecology, 4,* 36–41.
6. Granovsky., A. A. PC GAMESS version 7.1.5 (http://classic.chem.msu.su/gran/gamess/index.html).
7. Bürgi, T., Droz, T., & Leutwyler, S. (1995). *Chemical Physics Letters, 246,* 291–299.
8. Matsumoto, Y., Ebata, T., & Mikami, N. J. (1998). *Chemical Physics Letters, 109,* 6303–6311.

9. Schemmel, D., & Schütz, M. J. (2008). *Chemical Physics Letters, 129,* 34301.

10. Famulari, A., Raimondi, M., Sironi, M., & Gianinetti, E. (1998). *Chemical Physics Letters, 232,* 275–287.

11. Hedberg, K., Hedberg, L., & Bethune, D. S. et al. (1991). *Science, 254,* 410–412.

12. Copley, J. R. D., Neumann, D. A., & Cappelletti, R. L., & Kamitakahara, W. A. (1992). *Journal of Physics and Chemistry of Solids, 53,* 1353–1371.

13. Lipanov, A. M., Kodolov, V. I., Khokhriakov, N. V., Didik, A. A., Kodolova, V. V., & Semakina, N. V. (2005). *Alternative power engineering and ecology, 2,* 58–63.

14. Valiev M., Bylaska E. J., Govind, N., Kowalski, K., Straatsma, T. P., van Dam, H. J. J., Wang, D., Nieplocha, J., Apra, E., Windus T. L., & de Jong W.A. (2010). *Computer Physics Communication, 181*(9), 1477–1489.

15. Famulari, A., Raimondi, M., Sironi, M., & Gianinetti, E. (1998). *Chemical Physics Letters, 232,* 289–298.

16. Spectral database for organic compounds AIST. http://riodb01.ibase.aist.go.jp/sdbs/cgi-bin/cre_index.cgi?lang=eng.

17. HyperChem 8 manual. (2002). *Computational Chemistry* (p. 149, 369).

18. Seeger D. M., Korzeniewski C., & Kowalchyk W. (1991). *Journal of Physical Chemistry, 95,* 68–71.

19. Kazicina L. A., & Kupletskaya N. B. (1971). *Application of UV-, IR- and NMR spectroscopy in organic chemistry* (p. 264). Moscow: Chemistry.

20. Nakamoto K. (1966). *Infrared spectra of inorganic and coordination compounds* (p. 412). Moscow: Chemistry.

21. Kodolov V. I., & Khokhriakov N. V. (2009). *Chemical physics of formation and transformation processes of nanostructures and nanosystems* (Vol. 1 p. 365, Vol. 2 p. 4153). Izhevsk: Izhevsk State Agricultural Academy.

22. Kodolov V. I., Khokhriakov N. V., Trineeva V. V., & Blagodatskikh I. I. (2008). Activity of nanostructures and its expression in nanoreactors of polymeric matrixes and active media. *Chemical Physics and Mesoscopy, 10*(4), 448–460.

23. Panfilov B. F. (2010). Composite materials: Production, application, market tendencies. *Polymeric Materials, 2–3,* 40–43.

24. Bulgakov V. K., Kodolov V. I., & Lipanov A. M. (1990). Modeling of polymeric materials combustion (p. 238). Moscow: Chemistry.

25. Shuklin S. G., Kodolov V. I., Larionov K. I., & Tyurin S. A. (1995). Physical and chemical processes in modified two-layer fire- and heat-resistant epoxy-polymers under the action of fire sources. *Physics of Combustion and Explosion, 31*(2), 73–79.

26. Glebova, N. V. & Nechitalov, A. A. (2010). Surface functionalization of multi-wall carbon nanotubes. *Journal of Technical Physics, 36*(19), 12–15.

27. Patent 2393110 Russia Technique of obtaining carbon metal containing nanostructures. Kodolov V. I., Vasilchenko Yu. M., Akhmetshina L. F., Shklyaeva D. A., Trineeva V. V., Sharipova A. G., Volkova E. G., Ulyanov A. L., Kovyazina O. A.; declared on 17.10.2008, published on 27.06.2010.

28. Patent 2337062 Russia Technique of obtaining carbon nanostructures from organic compounds and metal containing substances. Kodolov V. I., Kodolova V. V. (Trineeva), Semakina N. V., Yakovlev G. I., Volkova E. G. et al; declared on 28.08.2006, published on 27.10.2008.

29. Kodolov V. I., Trineeva V. V., Kovyazina O. A., & Vasilchen-ko Yu. M. (2012). Production and application of metal / carbon nano-composites. *The problems of nanochemistry for the creation of new materials* (pp. 17–22). Torun: IEPMD.
30. Kodolov, V. I., & Trineeva, V. V (2012). Perspectives of idea development about nano-systems self-organization in polymeric matrixes. *The problems of nanochemistry for the creation of new materials* (pp. 75–100). Torun: IEPMD.
31. Schmidt, M. W., Baldridge, K. K., & Boatz, J. A. et al. (1993). *Journal of Computational Chemistry, 14,* 1347–1363.

CHAPTER 6

APPLICATION OF GENETIC ENGINEERING: PART I

A. S. KOROVASHKINA, S. V. KVACH, and A. I. ZINCHENKO

CONTENTS

6.1   Cloning of Multi-Copy C_pG Motifs In Bacterial Plasmids 115
    6.1.1   Introduction 115
    6.1.2   Materials and Methods 116
    6.1.3   Results and Discussion 118
    6.1.4   Conclusion 122
Keywords 122
References 122

6.1   CLONING OF MULTI-COPY C$_p$G MOTIFS IN BACTERIAL PLASMIDS

6.1.1   INTRODUCTION

Two plasmids with different copy number of unmethylated CpG dinucleo-tides in particular sequence contexts (CpG motifs) were constructed. The plasmid pCpG-KH11 carries 96 copies of CpG motif GTCGTT, with latter being the most effective in stimulation of the human immune system in accordance with literature data. The plasmid pCpG-KM11 contains 120 repetitions of GACGTT-motif found to enhance immune effects in mice, fish, and so on. The sequence of inserted fragments was verified by PCR and restriction enzyme digestion.

It is suggested that constructed plasmids may be used to improve pro-tein and DNA vaccines and to treat cancer, infectious and allergic diseases.

Unmethylated CpG dinucleotides in particular sequence contexts (CpG motifs), which are present in bacterial (but not mammalian) DNA stimulate innate host defense mechanisms by activating Toll-like receptor 9 (TLR9) signaling pathways [1–3]. Synthetic oligodeoxynucleotides (ODNs) con-taining CpG motifs have been shown to act as potent adjuvants that can induce T helper-1 (Th-1) type immune responses by mimicking the effects of bacterial CpG DNA [4–7]. Synthetic CpG ODNs can improve immune responses to a variety of coadministered antigens, and therefore represent a promising new approach for improving vaccine efficacy [8–12]. Recent-ly, synthetic CpG ODNs have been shown to enhance the immunogenicity of various microbial and tumor antigens in humans [13–16].

ODNs with natural phosphodiester bonds (PO-ODNs) are susceptible to nuclease degradation in cells [17]. Therefore, in order to provide resis-tance to nuclease activity and ensure efficient uptake of CpG ODNs by cells, phosphorothioate backbone-modified CpG ODNs containing sulfur substituted for the non bridging oxygen atoms (PS-ODNs) were devel-oped [18, 19]. However, several laboratories have shown that severe side effects occurred in PS-ODNs-treated mice [20–23]. Therefore, production of a potent immunostimulatory CpG PO-ODN that does not cause side ef-fects is highly desirable for inducing a well-controlled immune response.

Another way of CpG DNA production is construction of CpG-enriched plasmids with subsequent cloning in *Escherichia coli* cells. It is known that CpG motifs present in the backbone of DNA vaccines play an important role in the development of immune responses, and increasing the number of the motifs can enhance immune responses to various antigens [24-28]. Compared with synthetic ODNs, plasmids have more stable chemical properties and are economical in scaled-up production. Moreover, such CpG-containing plasmids are non toxic for organisms as they lack unnatural phosphorothioate linkages. Hence, CpG-enriched plasmids may prove to be more useful than protein or peptide vaccine adjuvants.

Taking all aspects into consideration, the aim of the research was construction of recombinant CpG-enriched plasmids on the basis of a multi-copy vector.

6.1.2 MATERIALS AND METHODS

To construct multi-copy CpG fragments, two partially overlapping primers containing four CpG motifs were chemically synthesized (CpG-F and CpG-R). *Hind*III restriction sites were inserted at 5' ends of primers (the n*ucleobases* are in italic).

Primers for cloning of **GTCGTT**-motif:

CpG-F-human—

5'-*AGCTT*CGTCGTTTTGTCGTTTTGTCGTTTTGTCGTT*A*-3';

CpG-R-human—

5'-*AGCTT*AACGACAAAACGACAAAACGACAAAACGAC*GA*-3'.

Primers for cloning of **GACGTT**-motif:

CpG-F-mouse—

5'-*AGCTT*CGACGTTTTGACGTTTTGACGTTTTGACGTT*A*-3';

CpG-R-mouse—

5'-*AGCTT*AACGTCAAAACGTCAAAACGTCAAAACGTC*GA*-3'.

To anneal the primers, $50\,\mu L$ of the reaction mixture containing 500pmol of each primer, 20mM Tris-HCl (pH 7.5), 10mM $MgCl_2$ and 50mM NaCl was heated at 95°C and slowly cooled down to 4°C. The products obtained were double-stranded DNA molecules with sticky 3' and 5' complementary ends. After phosphorylation of 5' ends with T4 polynucleotide kinase (Sileks, Russia), the products were ligated to each other. The fragments were separated using agarose gel electrophoresis and the ones with the highest size (900–1,300 base pairs), where isolated and purified with Min-Elute Gel Extraction Kit (Qiagen, Germany) according to manufacturer recommendations.

Purified double-stranded CpG polynucleotides were ligated into the vector pXcmkn12 (Cloning Vector Collection, Japan) previously digested with *Hind*III. A 10μL aliquot of the ligation mixture was used to transform *E. coli* DH5α competent cells (Novagen, USA). Cells were grown on Luria-Bertani (LB) agar plates with 100μg/mL ampicillin.

Several obtained colonies were screened by polymerase chain reaction (PCR) for plasmid pXcmkn-12 enriched with CpG motifs using primers terminating the cloned fragments. Amplification by PCR was performed in a reaction mixture (30μL) containing crude cell lysate as a DNA template, 5pmol of each synthetic primer (5'-GTAAAACGACGGCCAGT-3' and 5'-CAGGAAACAGCTATGAC-3'), four deoxyribonucleoside triphosphates (0.02mM each), $MgCl_2$ (2mM), $(NH_4)_2SO_4$ (17mM), 67mM Tris-HCl buffer (pH 8.3), 0.02% Tween 20, and 1U of *Taq* DNA polymerase (Promega, USA). The reaction was conducted for 30 cycles (30s at 95°C, 30s at 55°C and 1min at 72°C). Amplification products were analyzed using agarose gel electrophoresis.

Plasmids with the longest inserted fragments (designated pCpG-KH11 and pCpG-KM11) were chosen for subsequent manipulations. *E. coli* DH5α cells containing target plasmids were grown at 37°C with orbital shaking at 200rpm in 250mL Erlenmeyer flasks with 50mL of ZYM505 culture medium: tryptone—1.0%; yeast extract—0.5%; glycerol—0.5%; glucose—0.05%; Na_2HPO_4—0.025M; KH_2PO_4—0.025M; NH_4Cl—0.05M; Na_2SO_4—0.005M; ampicillin—150μg/mL in deionized water adjusted to pH 7.0 with KOH [29]. After 16hr of growth cells harvested by centrifugation for 10min at 12,000g, washed once with TE buffer (10mM Tris-HCl + 1mM EDTA; pH 8.0) and slurried in the same buffer. Plasmids were isolated from obtained cell biomass using alkaline lysis method [30].

Restriction analysis of constructed plasmids was performed by their incubation with mixture of restriction nucleases *Nde*I and *Eco*RI. The presence of multi-copy CpG motifs in cloned fragments was affirmed by incubation of restriction products with *Hind*III restrictase.

The obtained plasmids pCpG-KH11 and pCpG-KM11 were introduced into *E. coli* strain BLR(DE3). The transformed cells were grown at 37°C with orbital shaking at 200rpm in 250mL Erlenmeyer flasks containing 50mL of ZYM505 culture medium described above. Cultivation continued during 14–16hr. The obtained biomass was used for plasmid isolation by alkaline lysis method.

6.1.3   RESULTS AND DISCUSSION

Many new vaccines under development consist of rationally designed recombinant proteins that are relatively poor immunogens unless combined with potent adjuvants. There is only one adjuvant in common use in the U.S., aluminum phosphate or hydroxide (e.g. alum). This adjuvant, however, has significant limitations, particularly regarding generation of strong cell-mediated (T-cell) immune responses. In recent years, a novel adjuvant, JVRS-100, composed of cationic liposome DNA complexes (CLDC) has been evaluated for immune enhancing activity. The JVRS-100 adjuvant has been shown to elicit robust immune responses compared to CpG ODN and alum adjuvants, and it efficiently enhances both humoral and cellular immune responses [31]. Guan et al. [32] and Guo et al. [33] demonstrated that plasmids containing CpG motifs could function as immune potentiators to porcine reproductive and respiratory syndrome killed virus vaccines. In view of these scientific data, the aim of the work was preparation of recombinant CpG-enriched plasmids for further priority use as vaccine adjuvants.

Our work resulted in construction of two plasmids containing multi-copy CpG motifs. The plasmid pCpG-KH11 carries multiple repetitions of CpG motif GTCGTT. The cloned motif is known to be the most effective in stimulation of the human immune system. The plasmid pCpG-KM11 contains a series of GACGTT motif found to enhance immune effects in mice and other small-size animals. Schematic illustration of the different steps involved in the plasmids construction is shown in Figure 6.1.

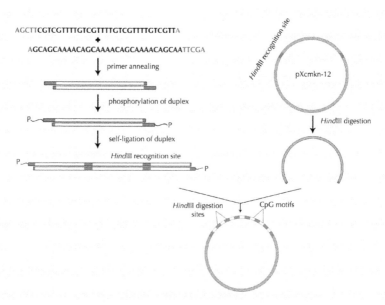

FIGURE 6.1   Schematic illustration of the different steps involved in the construction of the recombinant plasmid pCpG-KH11 containing multi-copy CpG motifs

Characterization of constructed plasmids was performed by restriction analysis. Purified plasmids were incubated with mixture of restriction endonucleases *Nde*I and *Eco*RI where recognition sites terminated the cloned fragments. The length of cloned fragments according to electrophoregram (Figure 6.1) was 960 base pairs (for GTCGTT motif) and 1,200 base pairs (for GACGTT motif).

Multi-copy CpG fragment consists of tandem repetitions containing 4 reiterations of CpG-motifs separated by *Hind*III restriction sites. To confirm structure of cloned fragments digestion products were additionally incubated with *Hind*III restrictase. Scheme of restriction analysis is shown in Figure 6.2. Electrophoregram of pCpG-KH11 and pCpG-KM11 restriction analysis is shown in Figure 6.3.

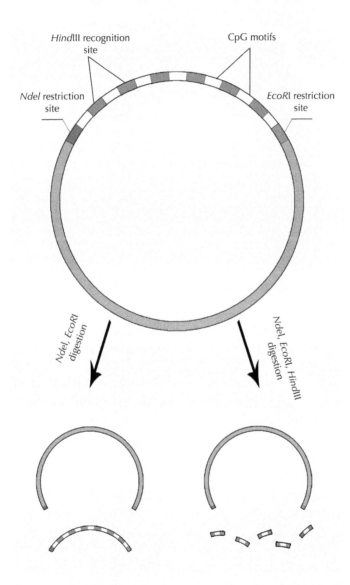

FIGURE 6.2 Scheme of restriction analysis of the plasmids containing multi-copy CpG motifs

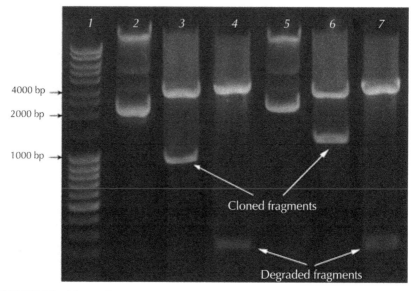

FIGURE 6.3 Electrophoregram of restriction analysis of constructed plasmids. *Line 1* is DNA molecular size markers; *line 2* is purified plasmid pCpG-KH11; *line 3* is pCpG-KH11 digestion products after treatment with *Nde*I and *Eco*RI; *line 4* is pCpG-KH11 digestion products after treatment with *Nde*I, *Eco*RI, and *Hind*III; *line 5* is purified plasmid pCpG-KM11; *line 6* is pCpG-KM11 digestion products after treatment with *Nde*I and *Eco*RI; *line 7* is pCpG-KM11 digestion products after treatment with *Nde*I, *Eco*RI, and *Hind*III

The number of cloned CpG motifs was calculated by the following formula:

$$X = \frac{D1}{40} \times 4,$$

where:

- $D1$—length of the cloned fragment identified after *Nde*I and *Eco*RI digestion (bp);
- 40—length of 1 tandem repetition of CpG motifs (bp);
- 4—number of CpG motifs in 1 tandem repetition.

By our estimates, the plasmid pCpG-KH11 carries 96 copies of GTCGTT motif and the plasmid pCpG-KM11 contains 120 repetitions of GACGTT motif.

The resultant recombinant plasmids pCpG-KH11 and pCpG-KM11 were introduced into *E. coli* strain BLR(DE3). The transformed cells were grown at 37°C with orbital shaking at 200rpm in 250mL Erlenmeyer flasks containing 50mL of ZYM505 culture medium. The recovered cell biomass was used for isolation and purification of target plasmids. It was shown that the plasmids were produced at a relatively high level (5.5–6.0mg/L), so that the selected recombinant strains can be successfully used for bio-technological production of CpG-enriched plasmids.

6.1.4 CONCLUSION

The completed work resulted in construction of the plasmids pCpG-KH11 carrying 96 copies of CpG motif GTCGTT and pCpG-KM11 with 120 repetitions of GACGTT motif. The cloned motifs are considered to be the most effective in stimulation of the human and mice immune system, respectively. Using methods of genetic engineering, strain *E. coli* CpG-KH11 producing plasmid pCpG-KH-11 and strain *E. coli* NBM-11 generating plasmid pCpG-KM11 have been obtained. The strains are potent sources of CpG-enriched plasmids and they make it possible to produce 5.5–6.0mg of target plasmid DNA per L of culture medium.

It is suggested that the constructed strains may be used in biotechnological production of pharmaceutically valuable CpG DNA.

KEYWORDS

- **CpG motifs**
- **Cationic liposome DNA complex**
- **ElectrophoregramPolymerase chain reaction**
- **Recombinant CpG-enriched plasmid**

REFERENCES

1. Krieg, A. M., Yi, A. K., Matson, S., Waldschmidt, T. J., Bishop, G. A., Teasdale, R., Koretzky, G. A., & Klinman, D. M. (1995). *Nature, 374*(6522), 546.

2. Hemmi, H., Takeuchi, O., Kawai, T., Kaisho, T., Sato, S., Sanjo, H., Matsumoto, M., Hoshino, K., Wagner, H., Takeda, K., & Akira S. (2000). *Nature, 408*(6813), 740.

3. Murad, Y. M. & Clay, T. M. (2009). *BioDrugs, 23*(6), 361.

4. Lipford, G. B., Bauer, M., Blank, C., Reiter, R., Wagner, H., & Heeg, K. (1997). *European Journal of Immunology, 27*(9), 2340.

5. Hoft, D. F., Eickhoff, C. S., Giddings, O. K., Vasconcelos, J. R. C., & Rodrigues, M. M. (2007). *Journal of Immunology, 179*(10), 6889.

6. Conforti, V. A., De Avila, D. M., Cummings, N. S., Zanella, R., Wells, K. J., Ulker, H., & Reeves, J. J. (2008). *Vaccine, 26*(7), 907.

7. Goldfarb, Y., Benish, M., Rosenne, E., Melamed, R., Levi, B., Glasner, A., & Ben-Eliyahu, S. (2009). Journal of Immunotherapy, 32(3), 280.8. Verthelyi, D., Kenney, R. T., Seder, R. A., Gam, A. A., Friedag, B., & Klinman, D. M. (2002). Journal of Immunology, 168, 1659.

9. Cheng, C., Bettahi, I., Cruz-Fisher, M. I., Pal, S., Jain, P., Jia, Z., Holmgren, J., Harandi, A. M., & De La Maza, L. M. (2009). *Vaccine, 27*(44), 6239.

10. Huang, M. H., Lin, S. C., Hsiao, C. H., Chao, H. J., & Yang, H. R. et al. (2010). *PLoS ONE, 5*(8), e12279.

11. Steinhagen, F., Kinjo, T., Bode, C., & Klinman, D. M. (2011). *Vaccine, 29*(17), 3341.

12. Rappuoli, R., Mandl, C. W., Black, S., & De Gregorio, E. (2011). *Nature Reviews Immunology, 11*(12), 865.

13. Cooper, C. L., Angel, J. B., Seguin, I., Davis, H. L., & Cameron, D. W. (2008). *Clinical Infectious Diseases, 46*(8), 1310.

14. Qian, F., Rausch, K. M., Muratova, O., Zhou, H., Song, G., & Diouf, A. (2008). *Vaccine, 26*(20), 2521.

15. Ellis, R. D., Mullen, G. E., Pierce, M., Ellis, R. D., Mullen, G. E., Pierce, M., Martin, L. B., Miura, K., Fay, M. P., Long, C. A., Shaffer, D., Saul, A., Miller, L.H., & Durbin, A. P. (2009). *Vaccine, 27*(31), 4104.

16. Karbach, J., Gnjatic, S., Bender, A., Neumann, A., Weidmann, E., Yuan, J., Ferrara, C. A., Hoffmann, E., Old, L. J., Altorki, N. K., & Jager, E. (2010). *International Journal of Cancer, 126*(4), 909.

17. Zhao, Q., Matson, S., Herrera, C. J., Fisher, E., Yu, H., & Krieg, A. M. (1993). *Antisense Research & Development, 3,* 53.

18. Stein, C. A., Subasinghe, C., Shinozuka, K., & Cohen, J. S. (1988). *Nucleic Acid Research, 16,* 3209.

19. Zhao, Q., Waldschmidt, T., Fisher, E., Herrera, C. J., & Krieg, A. M. (1994). *Blood, 84,* 3660.

20. Sparwasser, T., Hultner, L., Koch, E. S., Luz, A, Lipford, G. B., & Wagner, H. (1999). *Journal of Immunology, 162,* 2368.

21. Deng, G. M., Nilsson, I. M., Verdrengh, M., Collins, L. V., & Tarkowski, A. (1999). *Nature Medicine, 5,* 702.

22. Heikenwalder, M., Polymenidou, M., Junt, T., Sigurdson, C., Wagner, H., Akira, S., Zinkernagel, R., & Aguzzi, A. (2004). *Nature Medicine, 10,* 187.

23. Kim, D., Rhee, J. W., Kwon, S., Sohn, W. J., Lee, Y., Kim, D. W., Kim, D. S., & Kwon, H. J. (2009). *Biochemical and Biophysical Research Communications, 379,* 362.

24. Kojima, Y., Xin, K. Q., Ooki, T., Hamajima, K., Oikawa, T., Shinoda, K., Ozaki, T., Hoshino, Y., Jounai, N., Nakazawa, M., Klinman, D., & Okuda, K. (2002). *Vaccine, 20*(23–24), 2857.
25. Pontarollo, R. A., Babiuk, L. A., Hecker, R., & Van Drunen Littel-Van Den Hurk, S. (2002). *Journal of General Virology, 83,* 2973.
26. Coban, C., Ishii, K. J., Gursel, M., Klinman, D. M., & Kumar, N. (2005). *Journal of Leukocyte Biology, 78*(3), 647.
27. Liu, L., Zhou, X., Liu, H., Xiang, L., & Yuan, Z. (2005). *Immunology, 115*(2), 223.
28. Chen, Z., Cao, J., Liao, X., Ke, J., Zhu, S., Zhao, P., & Qi, Z. (2011). *Viral Immunology, 24*(3), 199.
29. Studier, F. W. (2005). *Protein Expression and Purification, 41*(1), 207.
30. Sambrook, J., Fritsch, E. F., & Maniatis, T. (1989). *Molecular cloning: A laboratory manual* (p. 2222). New York: Cold Spring Harbor Laboratory Press.
31. Chang, S., Warner, J., Liang, L., & Fairman, J. (2009). *Biologicals, 37*(3), 141.
32. Quan, Z., Qin, Z. G., Zhen, W., Feng, X. Z., Hong, J., & Fei, Z. H. (2010). *Veterinary Immunology and Immunopathology, 136*(3–4), 257.
33. Guo, X., Zhang, Q., Hou, S., Zhai, G., Zhu, H., & Sanchez-Vizcaino, J. M. (2011). *Veterinay Immunology and Immunopathology, 144*(3–4), 405.

CHAPTER 7

APPLICATION OF GENETIC ENGINEERING: PART II

A. S. KOROVASHKINA, S. V. KVACH, M. V. KVACH, A. N. RYMKO, and A. I. ZINCHENKO

CONTENTS

7.1   Enzymatic Synthesis Of 3', 5'-Cyclic Diguanylate Using Full-Length Diguanylate Cyclase of Thermotoga Maritima 126

    7.1.1   Introduction 126

    7.1.2   Materials and Methods 128

    7.1.3   Results and Discussion 130

    7.1.4   Conclusion 134

Keywords 134

References 134

7.1   ENZYMATIC SYNTHESIS OF 3', 5'-CYCLIC DIGUANYLATE USING FULL-LENGTH DIGUANYLATE CYCLASE OF *THERMOTOGA MARITIMA*

7.1.1   INTRODUCTION

Using methods of genetic engineering, two strains producing recombinant full-length diguanylate cyclase (DGC) of *Thermotoga maritima* with native and mutant allosteric sites have been constructed. The obtained strains have several-fold higher DGC-producing level than those described in literature. It has been shown that target enzymes are produced substantially in the form of active inclusion bodies. Seven-fold increase in activity of the *T. maritima* DGC was caused by mutation in allosteric site. Possibility of applying full-length DGC of *T. maritima* in the form of inclusion bodies for synthesis of 3', 5'-cyclic diguanylate has been originally demonstrated.

The 3', 5'-cyclic diguanylate (c-di-GMP) is the second messenger found in bacteria for intracellular signal transduction [1]. Accumulating evidence suggests that c-di-GMP regulates a wide variety of bacterial processes such as biofilm formation, motility, and virulence expression [2]. Moreover, recent studies have demonstrated that exogenous c-di-GMP treatment inhibits adhesive *Staphylococcus aureus* cell-to-cell interactions and biofilm formation [3] and that it is effective also in a murine model of mastitis infection [4]. These findings emphasize its prominent antimicrobial prospects. It was demonstrated that c-di-GMP stimulates the immune system to prevent bacterial infections and is therefore evaluated as a potential vaccine adjuvant [5, 6]. It was also reported that c-di-GMP inhibits cancer cell proliferation *in vitro* and could therefore be used as an antitumor therapeutic agent [7].

Characterization of the macromolecules involved in c-di-GMP signaling, including the enzymes and receptors, implies the use of c-di-GMP as substrate or ligand for biochemical assay. The c-di-GMP can also serve as starting material for the synthesis of c-di-GMP analogues to be used as chemical probes or inhibitors [8]. It appears therefore, that large quantities of this cyclic dinucleotide are required. Currently, c-di-GMP is mainly produced by chemical methods suffering from such limitations as complexity, high cost, and low yield of the end product [9]. On the other hand, the enzymatic production of c-di-GMP using diguanylate cyclase (DGC) domain-containing proteins involves only a single condensation of two guanosine-5′-triphosphate (GTP) molecules [10]. Reaction scheme is shown in Figure 7.1.

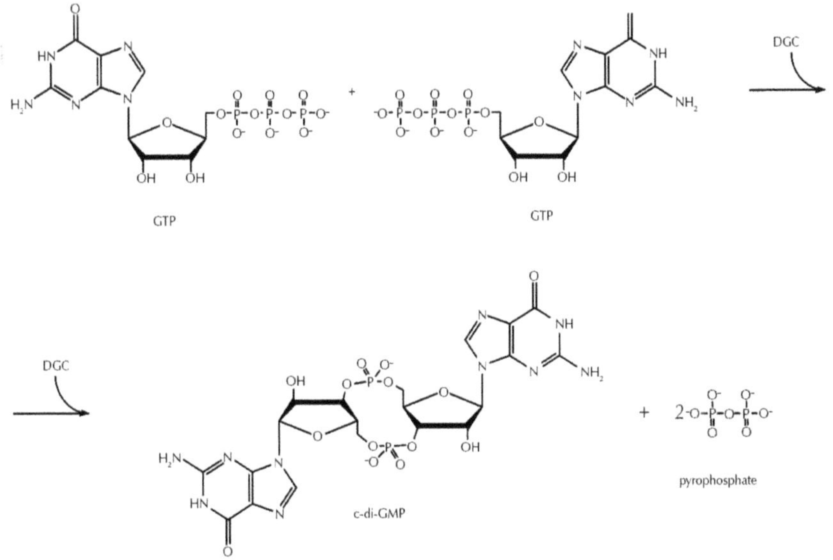

FIGURE 7.1   Synthesis of of c-di-GMP from GTP by DGC

Enzymatic large-scale synthesis of c-di-GMP from GTP has been reported for DGC of *Gluconacetobacter xylinum* [11], DGC of *Escherichia coli* [12], and a DGC fragment of *Thermotoga maritima* [8]. Mutations were introduced in allosteric sites of DGC of *Gluconacetobacter xylinum* and *T. maritima* to increase yield of target product. However, these proteins are produced at relatively low level (10–15mg/L of culture broth) or require complex and labor-consuming procedure of refolding. Furthermore, the mesophilic proteins of *Gluconacetobacter xylinum* and *E. coli* are rather unstable in solution.

Full-length DGC of *T. maritima* (TM1788, GenBank) comprises catalytic GGDEF domain and hydrophobic domain of unknown function [8]. It was shown earlier that the synthesis of c-di-GMP from two molecules of GTP requires the DGC protein to function in its dimeric form [1]. Previously used DGC fragment of *T. maritima* containing only catalytic domain in the reaction mixture corresponds to monomeric form and synthesis of c-di-GMP is achieved by the formation of transient dimer by the protein in the solution [8].

Considering the above mentioned aspects, we presumed that application of full-length DGC from *T. maritima* comprising both catalytic and hydrophobic domains would increase the reaction rate due to formation of stable DGC dimers.

The aim of the research was construction and characterization of *E. coli* strains producing full-length DGC of *T. maritima* with native and mutant allosteric sites and verification of c-di-GMP synthesis using the obtained enzymes.

7.1.2 MATERIALS AND METHODS

The DGC-encoding gene (*TM1788*) was amplified by PCR from the chromosomal DNA of *T. maritima* using the following synthetic oligonucleotide primers: DgcfullF 5′-ATTGTG*CATATG*AAAGTTTCCGGGGGGGGAA-GT-3′ and DgcfullR 5′-ATCTAG*CTCGAG*TGAGAGACTGAAATAA-GGCAC-3′. *Nde*I (DgcfullF) and *Xho*I (DgcfullR) restriction sites were inserted at 5′ ends of the primers (the bases are in italic). Point mutation R158A was introduced by megaprimer method [13] with primers: RXXD-f 5′- ACAGGGTT**GCA**AGAAGCGATGTGGTGGC-3′ and RXXD-r 5′-ATCGCTTCT**TGC**AACCCTGTCCAGAATCC-3′ (mutation is in bold).

Amplification was performed using 1U of High-Fidelity DNA Polymerase (Fermentas, Lithuania) and a PTC-200 GeneAmp PCR system (Bio-Rad, USA). The resulting products were isolated with Wizard SV gel and PCR clean-up system (Promega, USA), digested with *Nde*I and *Xho*I restriction endonucleases and ligated into the pET24b+ (Novagen, USA) expression vector, which had been restricted with same enzymes.

As a result, plasmids bearing the gene of full-length DGC of *T. maritima* with native (pETdgc-w7) and mutated (pETdgcR158A) allosteric sites were constructed. The recombinant plasmids were introduced into *E. coli* strain BL21(DE3). The transformed cells were grown at 37°C with orbital shaking at 200rpm in 250mL Erlenmeyer flasks containing 50mL of ZYM505 culture medium [14]: tryptone—1.0%; yeast extract—0.5%; glycerol—0.5%; glucose—0.05%; α-D-lactose—0.2%; Na_2HPO_4—0.025M; KH_2PO_4—0.025M; NH_4Cl—0.05M; Na_2SO_4—0.005M; kanamycin—50μg/mL in deionized water adjusted to pH 7.0 with KOH. Protein synthesis was triggered by autoinduction. Cells were harvested by centrifugation for 10min at 12,000g, washed once with buffer A containing 50mM Tris-HCl (pH 8.0) and 100mM KCl and then slurried in the same buffer. Cells sonication was performed using Sonifier-450 (Branson, USA) with following program: output—0.05kW; temperature—4°C; duration—600 pulses 0.5s each. Obtained crude cell lysate was centrifuged during 30min at 21,000g.

The precipitate containing inclusion bodies was washed with 1M urea solution containing 2% Triton X-100. After centrifugation, the precipitate was washed with buffer A to remove urea and Triton X-100 traces and resuspended in the same buffer.

Inclusion bodies concentration measurement was determined by modified Biuret method. In brief, suspension containing inclusion bodies was boiled during 5min with 1% sodium dodecyl sulfate (SDS). Further measurement was performed according to standard protocol [15].

Analysis of protein composition was conducted with SDS polyacrylamide gel electrophoresis (SDS-PAGE). Purity and molecular weight of obtained proteins was evaluated using program TotalLab TL120 (Nonlinear Dynamics, USA). DGC activity assay was carried out in 100μL reaction mixture containing 1μL freshly produced DGC, 1mM GTP, 50mM Tris-HCl (pH 8.0), 5mM $MgCl_2$, and 250mM NaCl. The reaction mixture was incubated at 55°C and aliquots at different time points were taken. The samples were loaded onto an TSK DEAE-3SW HPLC-column (LKB-Bromma, Sweden) and eluted at a flow rate 1mL/min by linear gradient from 3mM KCl + 3mM KH_2PO_4 (pH 3.3) to 360mM KCl + 360mM KH_2PO_4 (pH 4.9) for 25min. The amount of enzyme converting 1μmol of substrate per minute was assumed as 1 enzyme activity unit.

7.1.3 RESULTS AND DISCUSSION

Large-scale enzymatic synthesis of c-di-GMP using DGC of *Gluconace-tobacter xylinum* [11], *E. coli* [12], and DGC fragment of *T. maritima* [8] has previously been reported. The most promising from the described enzymes is the fragment of DGC from hyperthermophilic microorganism *T. maritima* due to its high thermostability [8]. However, this enzyme lacks hydrophobic domain existing in the full-length protein and therefore, can't form stable dimeric structure in the solution, which is necessary for c-di-GMP synthesis. We supposed that full-length DGC of *T. maritima* comprising both catalytic and hydrophobic domains would possess higher enzymatic activity due to formation of stable dimers in the reaction mixture.

Full-length DGC of *T. maritima* with native allosteric site (DGC^W) and DGC carrying allosteric site mutation R158A (DGC^{R158A}) were obtained in the current work to investigate the influence of allosteric inhibition on the enzyme activity.

At the initial stage, using methods of genetic engineering, two strains producing DGC of *T. maritima* with native (*E. coli* pTmadgc-w7) and mutant (*E. coli* pTmadgc) allosteric site have been constructed. Autoinduction of target proteins was performed in ZYM505 medium. Analysis of protein composition of cell lysate and its soluble and insoluble fractions was conducted with SDS-PAGE (Figure 7.2).

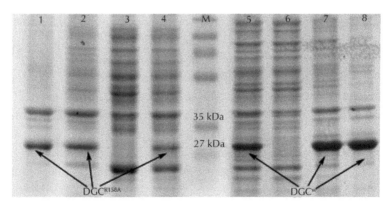

FIGURE 7.2 Electrophoregram of protein composition of E. coli pTmadgc and E. coli pTmadgc-w7 cells. Line *M*, protein molecular size markers; lines *4* and *5*, protein

compositions of *E. coli* pTmadgc и *E. coli* pTmadgc-w7 cells after autoinduction; lines *3* and *6*, protein compositions of soluble fraction of *E. coli* pTmadgc (*3*) and E. coli pTmadgc-w7 (*6*) cell lysate; lines *2* and *7*, protein compositions of insoluble fraction of *E. coli* pTmadgc (*2*) and *E. coli* pTmadgc-w7 (*7*) cell lysate; lines *1* and *8*, purified DGC[R158A] and DGC[w], respectively

The Figure 7.2 shows that the target proteins are produced mainly in the form of inclusion bodies. In the soluble fractions, the proteins with the molecular weight corresponding to full-length DGC (29kDa) were not detected. This could be explained by presence of hydrophobic domain at the N-terminus of full-length DGC leading to its irreversible aggregation [8].

Expression level of DGC[R158A] (6%) calculated with program TotalLab TL120 was approximately 4-fold lower than that of DGC[W] (23%). This might be due to higher enzymatic activity of DGC[R158A] and its increased toxicity for cells.

Earlier inclusion bodies were regarded as inactive protein aggregates and to obtain the soluble protein with natural activity, labor-consuming and inefficient procedure of refolding had to be arranged [16]. The recent studies have shown, however, that certain proteins possess enzymatic activity in the form of inclusion bodies [17, 18].

Taking it into account, we analyzed enzymatic activities of both soluble and insoluble fractions of cell lysate. No DGC-activity was revealed in the soluble fractions, while the activity values 4.23U/g protein for DGC[R158A] and 1.84U/g protein for DGC[W] were recorded in the insoluble fractions. However, GTP was cleaved to guanosine-5′-diphosphate and guanosine-5′-monophosphate due to presence of interfering phosphatase activity in the enzyme preparations. That required further purification of obtained proteins. For this purpose, the precipitate containing inclusion bodies was washed with 1M urea solution containing 2% Triton X-100. Purity of obtained protein preparations was determined using SDS-PAGE (Figure 7.2). The results of conducted purification steps are summarized in Table 7.1.

TABLE 7.1 Results of DGC^{R158A} and DGC^W purification

Fraction	DGC^{R158A}					DGC^W				
	Total activity (U)	Total protein (g)	Specific activity (U/g protein)	Relative purification factor	Recovery (%)	Total activity, U	Total protein (g)	Specific activity (U/g protein)	Relative purification factor	Recovery, %
Crude cell lysate	2	1.27	1.57	1	100	1.4	1.9	0.74	1	100
Precipitate of the cell lysate	2.3	0.55	4.23	2.7	115	1.9	1.032	1.84	2.5	135
Precipitate of the cell lysate after washing with 1M urea + 2% Triton X-100	4.8	0.26	18.68	12	240	3.6	0.825	4.4	5.9	259

There was approximately 25-fold increase in enzymatic activity in the protein preparations after purification procedure. It might be caused by presence of proteins with interfering phosphatase activity in the initial cell lysate. In the purified protein preparations, no phosphatase activity was detected. Purification grade of washed inclusion bodies estimated by program TotalLab TL120 was 31% for DGC^{R158A} and 50% for DGC^W. Thus, from 1L of culture liquid we obtained 80mg DGC^{R158A} and 410mg DGC^W with specific activities 60.25 and 8.8U/g of target protein, respectively. The specific activity of mutant DGC is 7-fold higher than that of native protein.

The obtained proteins were used for small-scale c-di-GMP synthesis. Reaction mixture containing 50mM Tris-HCl (pH 8.0), 250mM NaCl, 5mM MgCl$_2$, 1mM GTP and 2.5μM DGC^{R158A} or 9.5μM DGC^W was incubated at 55°C. Reaction progress was monitored using high-performance liquid chromatography (Figure 7.3).

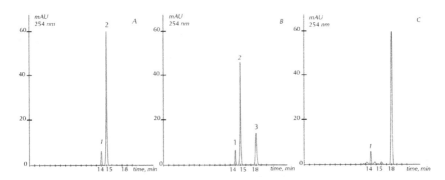

FIGURE 7.3 HPLC analyses of aliquots from the reaction mixture before (**a**) and after incubation with DGC for 20min (**b**) and 60min (**c**)

It was shown that under described conditions approximately 100% of 1mM GTP was converted into c-di-GMP during 1hr by 2.5μM DGC^{R158A} and 95% of the substrate at the same concentration was transformed into the end product in 5hr by 9.5μM DGC^W.

As reported earlier, under the same conditions, the reaction was completed in 1hr by 10μM fragment of mutated *T. maritima* DGC [8]. The latter enzyme level is 4-fold higher than the amount of the mutant full-length DGC of *T. maritima* used in the current work.

7.1.4 CONCLUSION

This work resulted in construction of strains *E. coli* pTmadgc-w7 and *E. coli* pTmadgc producing full-length recombinant DGC of *T. maritima* with native and mutant allosteric sites. The obtained strains have several-fold higher DGC producing capacity than those described in literature. It has been shown that target enzymes are produced substantially in the form of inclusion bodies that facilitate the protein purification process and subsequent isolation of the enzyme from the reaction mixture. Seven-fold increase in activity of the *T. maritima* DGC was triggered by mutation introduced into allosteric site. Moreover, full-length mutant enzyme DGC^{R158A} is more active than the truncated protein described previously by Rao et al. [8] and it allows to accomplish the synthesis of c-di-GMP with lower enzyme amounts probably due to presence of hydrophobic domain.

KEYWORDS

- **Cell lysate**
- **Diguanylate cyclase**
- **Guanosine-5'-triphosphate**
- **SDS polyacrylamide gel electrophoresis**
- **Thermotoga maritima**

REFERENCES

1. Romling, U., Gomelsky, M., & Galperin, M. Y. (2005). *Molecular Microbiology, 57*(3), 629.
2. Jenal, U., & Malone, J. (2006). *Annual Review Genetics, 40,* 385.

3. Karaolis, D. K. R., Rashid, M. H., Chythanya, R., Luo, W., Hyodo, M., & Hayakawa, Y. (2005). *Antimicrobial Agents Chemotherapy, 49*(3), 1029.
4. Brouillette, E., Hyodo, M., Hayakawa, Y., Karaolis, D. K., & Malouin, F. (2005). *Antimicrobial Agents Chemotherapy, 49*(8), 3109.
5. Karaolis, D. K. R., Means, T. K., Yang, D., Takahashi, M., Yoshimura, T., Muraille, E., Philpott, D., Schroeder, J. T., Hyodo, M., Hayakawa, Y., Talbot, B. G., Brouillette, E., & Malouin, F. (2007). *Journal of Immunology, 178*(4), 2171.
6. Ogunniyi, A. D., Paton, J. C., Kirby, A. C., Mccullers, J. A., Cook, J., Hyodo, M., Hayakawa, Y., & Karaolis, D. K. R. (2008). *Vaccine, 26*(36), 4676.
7. Karaolis, D. K. R., Cheng, K., Lipsky, M., Elnabawi, A., Catalano, J., & Hyodo, M. (2005). *Biochemical and Biophysical Research Communications, 329*(1), 40.
8. Rao, F., Pasunooti, S., Ng, Y., Zhuo, W., Lim, L., Liu, A. W., & Liang, Z. H. (2009). *Analytical Biochemistry, 389*(2), 138.
9. Kiburu, I., Shurer, A., Yan, L., & Sintim, H. O. (2008). *Molecular Biosystem, 4*(6), 518.
10. Chan, C., Paul, R., Samoray, D., Amiot, N. C., Giese, B., Jenal, U., & Schirmer, T. (2004). *Proceedings of the National Academy of Sciences USA, 101*(49), 17084.
11. Spehr, V., Warrass, R., Hocherl, K., & Ilg, T. (2011). *Applied Biochemistry and Biotechnology, 165*(3–4), 761.
12. Zahringer, F., Massa, C., & Schirmer, T. (2011). *Applied Biochemistry and Biotechnology, 163*(1), 71.
13. Mcpherson, M. J. & Moller, S. G. (2000). PCR: The basics from background to bench (p. 276). New York: BIOS.
14. Studier, F. W. (2005). *Protein Expression and Purification, 41*(1), 207.
15. Dawson, R. M. C., Eleiott, D. C., Eleiott, W. H., & Jones, K. M. (1991). Data for biochemical research (p. 544). Moscow: Mir (in Russian).
16. Rudolph, R., & Lilie, H. (1996). *The FASEB Journal, 10*(1), 49.
17. Nahalka, J., & Patoprsty, V. (2009). *Organic and Biomolecular Chemistry, 7*(9), 1778.
18. Garcia-Fruitos, E., Vazquez, E., Diez-Gil, C., Corchero, J. L., Seras-Franzoso, J., Ratera, I., Veciana, J., & Villaverde, A. (2012). *Trends in Biotechnology, 30*(2), 65.

Section 2: Production of Nanofibers

NANOPOLYMERS: FUNDAMENTALS AND BASIC CONCEPTS

A. K. HAGHI and A. HAMRANG

CONTENTS

8.1   Introduction 141

    8.1.1   Nanofibers 142

    8.1.2   Drawing 143

    8.1.3   Template Synthesis 144

    8.1.4   Phase Separation 145

    8.1.5   Self-Assembly 145

8.2   Electrospinning 145

8.3   Definitions 147

    8.3.1   Homopolymers 149

    8.3.2   Copolymers 149

    8.3.3   Synthetic Polymers 153

    8.3.4   Natural Polymers 154

    8.3.5   Proteins 155

8.4   Production of Nanofibers 157

    8.4.1   Processing Condition 158

8.5   Theory 159

    8.5.1   Lattice Boltzmann method (LBM) 164

    8.5.2   Hydrodynamics 165

8.5.3   Dimensionless Non Newtonian Fluid Mechanics
        Method... 168

8.5.4   Detection of X-ray Generated by Electrospinning 168

8.5.5   Modeling.. 169

8.6   Effect of Voltage And Spinning Distance on Morphology
      and Diameter.. 181

8.7   Concluding Remarks.. 183

Keywords 192

References... 192

8.1 INTRODUCTION

The term "nano" like many other prefixes used in conjunction with System International d'unites (SI units), "nano" comes from a language other than English. Originating from the Greek word "nannas" for "uncle", the Greek word "nanos" or "nannos" refer to "little old man" or "dwarf". Before the term nanotechnology was coined and became popular, the prefixes "nanno-" or "nano-" were used in equal frequency, although not always technically correct.

The use of nanostructure materials is not a recently discovered era. Back in the 4th century AD, the Romans used nano sized metal particles to decorate cups. The first known and the most famous example is the Lycurgus cup. In this cup as well as the famous stained glasses of the 10th, 11th, and 12th centuries, metal nanoparticles account for the visual appearance. This property is used in other ways in addition to stained glass. For example, particles of titanium dioxide (TiO_2) have been used for a long time as the sun-blocking agent in sunscreens.

Today, nanotechnology refers to technological study and application involving nanoparticles.

In general, nanotechnology can be understood as a technology of design, fabrication, and applications of nanostructures and nanomaterials. Nanotechnology also includes fundamental understanding of physical properties and phenomena of nanomaterials and nanostructures. Nanostructure is the study of objects having at least one dimension within the nano-scale. A nanoparticle can be considered as a zero-dimensional nano-element, which is the simplest form of nanostructure.

Since the turn of the century, the research and development on the nature-inspired manufacturing technology generally referred to as "biomimetics", have been coming to the fore in Europe and the United States. The term "biomimetics" (in Japan, the term is translated literally as "Imitation of Living Things") was proposed by German-American neurophysiologist Otto Schmitt in the latter half of the 1950s.

Scientists found that, on the surface of lotus leaves, the surface microstructure and secretion of wax-like compounds have a synergetic effect that produces super-hydrophobic property and is self-cleaning. The discovery of the lotus effect induced surface-structure research on living things that dwell in aquatic environments. The petals of roses, clivias, and sunflowers show superhydrophilicity, but at the same time they have so strong adsorption power to water that they hold water drops even if they are held upside down.

Other researches in biomimetics are: Adhesive materials inspired by gecko legs, swimsuit material inspired by sharkskin riblet, structural color materials inspired by butterflies and jewel beetles, anti-reflective materials inspired by the moth-eye structure, low friction materials inspired by sandfish, and sensor materials inspired by the sensing ability of insects.

There are two approaches to the synthesis of nanomaterials and the fabrication of nanostructures: Top-down and bottom-up, which depend upon the final size and shape of a nanostructure or nanodevice. Attrition or milling is a typical top-down method in making nanoparticles, whereas the colloidal dispersion is a good example of bottom-up approach in the synthesis of nanoparticles. Bottom-up approach is often emphasized in nanotechnology literature. Examples include the production of salt and nitrate in chemical industry, the growth of single crystals and deposition of films in electronic industry. The bottom-up approach refers to the approach to build a material up from the bottom: Atom-by-atom, molecule-by-molecule, or cluster-by-cluster. Although the bottom-up approach is nothing new, it plays an important role in the fabrication and processing of nanostructures and nanomaterials. When structures fall into a nanometer scale, there is little choice for a top-down approach. All the tools we have possessed are too big to deal with such tiny objects.

8.1.1   NANOFIBERS

Human beings have used fibers for centuries. In 5000 BC, our ancestors used natural fibers such as wool, cotton, silk, and animal fur for clothing. Mass production of fibers, dates back to the early stages of the industrial revolution. The first man-made fiber—viscose—was presented in 1889 at the World Exhibition in Paris. Traditional methods for polymer fiber production include melt spinning, dry spinning, wet spinning and gel-state spinning. These methods rely on mechanical forces to produce fibers by extruding a polymer melt or solution through a spinneret and subsequently drawing the resulting filaments as they solidify or coagulate. These methods allow the production of fiber diameters typically in the range of 5–500microns. At variance, electrospinning technology allows the production of fibers of much smaller dimensions. The fibers are produced by using an electrostatic field.

A nanofiber is a nanomaterial in view of its diameter, and can be considered a nanostructured material if filled with nanoparticles to form composite nanofibers. Also nanofibers are defined as fibers with diameter less than 1000nm. There are various techniques for polymeric nanofiber production such as drawing, template synthesis, phase separation, self-assembly, and electrospinning.

8.1.2 DRAWING

Drawing is a process whereby nanofibers are produced when a micropipette with a tip few micrometers in diameter is dipped into a droplet near the contact line using a micromanipulator and then withdrawn from the liquid at about $1 \times 10^{-4} m \cdot s^{-1}$. When the end of the micropipette touches the surface, the drawn nanofiber is then deposited (Figure 8.1). To draw fibers, a visco–elastic material is required so that it can endure the strong deformation and at the same time it has to be cohesive enough to support the stress that build up in the pulling process. The drawing process can be considered as dry spinning at a molecular level.

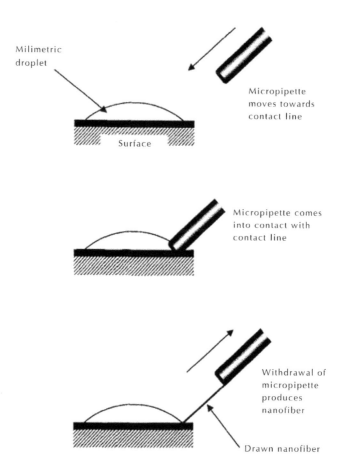

FIGURE 8.1   Production of nanofiber by drawing

This method can be used to make one-by-one and very long single nanofibers.

8.1.3 TEMPLATE SYNTHESIS

Template synthesis implies the use of a template or mold to obtain a desired material or structure. A nanoporous membrane is used as template to make nanofibers of solid (i.e., fibril) or hollow (i.e., tubule) shape. Using this method, nanometer tubules and fibrils of various raw materials such as electronically conducting polymers, metals, semiconductors, and carbons can be fabricated. This method cannot, however, make one-by-one continuous fibers.

8.1.4   PHASE SEPARATION

In this technique, porous polymer scaffolds are produced by the removal of solvent through freeze-drying or extraction. This process takes a longer time to transfer the solid polymer into the nanoporous foam. Phase separation can be induced by changing the temperature called the thermal induced phase separation or by adding non solvent-induced-phase separation.

8.1.5   SELF-ASSEMBLY

The self-assembly refers to the process in which individual, pre-existing components organize themselves into desired patterns and functions. However, this method like phase separation method is time consuming. Finally, it seems that the electrospinning is the only method which can be further developed to produce large quantity of one-by-one continuous nanofibers from synthetic polymers or natural polymeric materials.

However, it should be pointed out that the development of functional nanofibers via self-assembly method is an increasing area of interest. Many kinds of polymer nanostructures have been developed successfully by self-assembly.

8.2   ELECTROSPINNING

The term "electrospinning" has been derived from "electrostatic spinning". The innovative setup to produce fibers, dates back to 1934 when Formhals published a patent describing an experimental setup for the production of polymer filaments.

The richness of fiber structures and also of non woven architectures that can be accessed via electrospinning is extremely broad and it has become apparent that highly fine fibers with diameters down to just a few nanometers can be prepared via electrospinning.

A basic electrospinning setup, as shown in Figure 8.2 (a), consists of a container for polymer solution, a high-voltage power supply, spinneret (needle), and an electrode collector. During electrospinning, a high electric voltage is applied to the polymer solution and the electrode collector leading to the formation of a cone-shaped solution droplet at the tip of the spinneret, so called "Taylor cone". A solution jet is created when the voltage reaches a critical value, typically 5–20kV, at which the electrical forces overcome the surface tension of the polymer solution. Under the action of the high electric field, the polymer jet starts bending or whipping around stretching it thinner. Solvent evaporation from the jet results in dry/semidry fibres which randomly deposit onto the collector forming a non woven nanofiber web in most cases (Figure 8.2b).

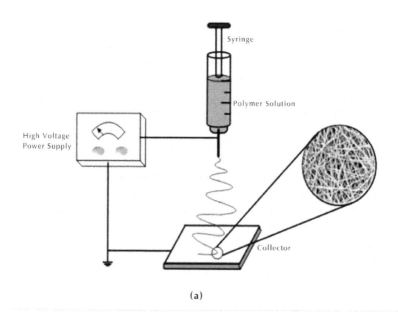

(a)

FIGURE 8.2 *(Continued)*

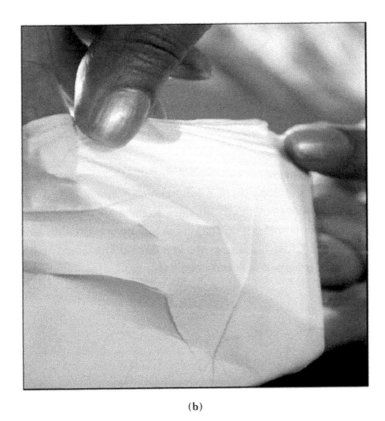

(b)

FIGURE 8.2   (a) Basic apparutus for electrospinning, (b) photo of typical electrospun nanofiber mat.

8.3   DEFINITIONS

Polymers (or macromolecules) are very large molecules made up of smaller units called monomers or repeating units covalently bonded together (Figure 8.3). This specific molecular structure (chainlike structure) of polymeric materials is responsible for their intriguing mechanical properties.

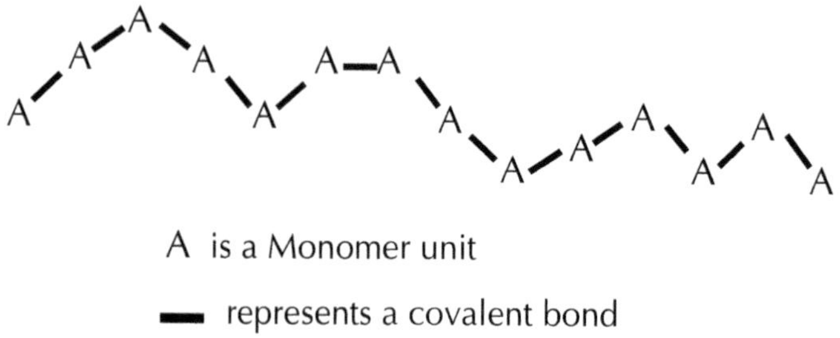

A is a Monomer unit

━ represents a covalent bond

FIGURE 8.3 Polymer chain.

A *linear polymer* consists of a long chain of monomers. A *branched polymer* has branches covalently attached to the main chain. *Cross-linked polymers* have monomers of one chain covalently bonded with monomers of another chain. Cross-linking results in a three-dimensional network; the whole polymer is a giant macromolecule (Figure 8.4).

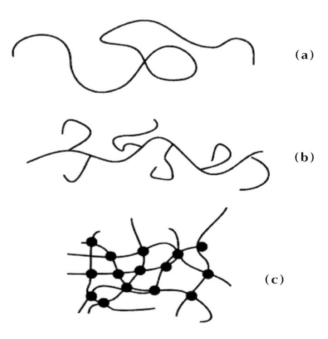

FIGURE 8.4 Types of molecular configuration: (a) linear chain, (b) branched molecule, and (c) cross-linked network: molecules are linked through covalent bonds; the network extends over the whole sample which is a giant macromolecule.

8.3.1 HOMOPOLYMERS

The formal definition of a homopolymer is a polymer derived from one species of polymer. However, the word homopolymer is often used broadly to describe polymers whose structure can be represented by multiple repetition of a single type of repeat unit which may contain one or more species of monomer unit. The latter is sometimes referred to as a structural unit.

8.3.2 COPOLYMERS

The formal definition of a copolymer is a polymer derived from more than one species of monomer. Copolymers have different repeating units. Furthermore, depending on the arrangement of the types of monomers in the polymer chain, we have the following classification:

- In *random* copolymers, two or more different repeating units are distributed randomly.
- *Alternating* copolymers are made up of alternating sequences of different monomers.
- In *block* copolymers, long sequences of a monomer are followed by longsequences of another monomer.
- *Graft* copolymers consist of a chain made from one type of monomers with branches of another type (Figure 8.5).

-A-A-A-A-A-A-A-A-A-A-A-A-A-A-A-A-A-A-A-A-A-A-A-A-A-A-A- (**a**)

-A-A-B-A-B-B-A-B-A-A-B-A-A-A-B-A-B-B-A-B-B-B-A-A-B-A-A- (**b**)

-A-B-A-B-A-B-A-B-A-B-A-B-A-B-A-B-A-B-A-B-A-B-A-B-A-B-A- (**c**)

FIGURE 8.5   (*Continued*)

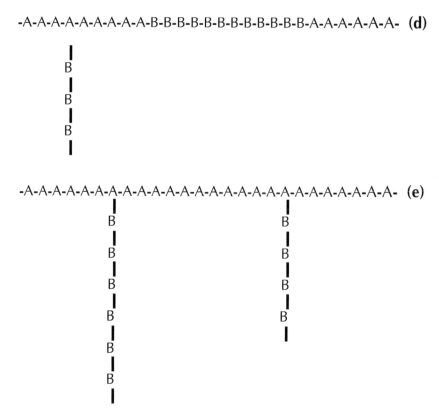

FIGURE 8.5   (a) Homopolymer, (b) random copolymer, (c) alternating copolymer, (d) Block copolymer, and (e) graft copolymer.

Another way to classify polymers is to adopt the approach of using their response to thermal treatment and to divide them into thermoplastics and thermosets. Thermoplastics are polymers which melt when heated and resolidify when cooled, while thermosets are those which do not melt when heated but at sufficiently high temperatures, decompose irreversibly.

In 1929, scientists proposed a generally useful differentiation between two broad classes of polymers: *Condensation polymers,* in which the molecular formula of the structural unit (or units) lacks certain atoms present in the monomer from which it is formed, or to which it may be degraded by chemical means, and *addition polymers*, in which the molecular formula of the structural unit (or units) is identical with that of the monomer from which the polymer is derived. Condensation polymers may be formed from monomers bearing two or more reactive groups of such character that they may condense inter-molecularly with the elimination of a byproduct, often water. For example, polyesters which are formed by condensation are shown in this reaction.

$$n \, \text{HO-R-OH} + \text{HOOC-R}^1\text{-COOH} \rightarrow \text{HO[-R-COO-R}^1\text{-COO-]}_n\text{H} + (n\text{-}1)\text{H}_2\text{O}$$

The most important class of addition polymers consists of those derived from unsaturated monomers such as vinyl compounds.

$$\text{CH}_2\text{=CHX} \rightarrow \text{CH}_2\text{-CHX-CH}_2\text{-CHX-}$$

In Tables 8.1 and 8.2 are some examples of addition and condensation polymers.

TABLE 8.1 Addition polymers

Polymer name	Monomer(s)	Polymer	Use
Polyethylene	$\text{CH}_2\text{=CH}_2$ (ethene)	$-\text{CH}_2-\text{CH}_2-$	Most common polymer. Used in bags, wire insulation, and squeeze bottles
Polypropylene	$\text{CH}_2\text{=CH}$ | CH_3 (1-propene)	$-\text{CH}_2-\text{CH}-$ | CH_3	Fibers, indoor-outdoor carpets, bottles
Polystyrene	$\text{CH}_2\text{=CH}$ (styrene)	$-\text{CH}_2-\text{CH}-$	Styrofoam, molded objects such as tableware (forks, knives and spoons), trays, videocassette cases.
Poly(vinyl chloride) (PVC)	$\text{CH}_2\text{=CH}$ | Cl (vinyl chloride)	$-\text{CH}_2-\text{CH}-$ | Cl	Clear food wrap, bottles, floor covering, synthetic leather, water and drain pipe
Polytetrafluoroethylene (Teflon)	$\text{CF}_2\text{=CF}_2$ (tetraflouroethene)	$-\text{CF}_2-\text{CF}_2-$	Nonstick surfaces, plumbing tape, chemical resistant containers and films
Poly(methyl methacrylate) (Lucite, Plexiglas)	CO_2CH_3 | $\text{CH}_2\text{=C}$ | CH_3 (methyl methacrylate)	CO_2CH_3 | $-\text{CH}_2-\text{C}-$ | CH_3	Glass replacement, paints, and household products
Polyacrylonitrile (Acrilan, Orlon, Creslan)	$\text{CH}_2\text{=CH}$ | CN (acrylonitrile)	$-\text{CH}_2-\text{CH}-$ | CN	Fibers used in knit shirts, sweaters, blankets, and carpets

TABLE 8.1   *Continued*

Poly(vinyl acetate) (PVA)	CH₂=CH | OOCCH₃ (vinyl acetate)	−CH₂−CH− | OOCCH₃	Adhesives (Elmer's glue), paints, textile coatings, and chewing gum
Natural rubber	CH₃ | CH₂=C−CH=CH₂ (2-methyl-1,3-butadiene)	CH₃ | −CH₂−C=CH−CH₂−	Rubber bands, gloves, tires, conveyor belts, and household materials
Polychlorprene (neoprene rubber)	Cl | CH₂=C−CH=CH₂ (2-methyl-1,3-butadiene)	Cl | −CH₂−C=CH−CH₂−	Oil and gasoline resistant rubber
Styrene butadiene rubber (SBR)	CH₂=CH (phenyl) CH₂=CH−CH=CH₂	−CH₂−CH−CH₂−CH−CH−CH₂− (phenyl)	Non-bounce rubber used in tires

TABLE 8.2   Condensation Polymers

Polymer name	Monomers	Polymer	Use
Polyamides (nylon)	HOC(CH₂)₄COH H₂N(CH₂)₆NH₂	−C(CH₂)₄C−NH(CH₂)₆NH−−	Fibers, molded objects
Polyesters (Dacron, Mylar, Fortrel)	HOC⟨⟩COH HO(CH₂)₂OH	−C⟨⟩C−O(CH₂)₂O−	Linear polyesters, fibers, recording tape
Polyesters (Glyptal resin)	(phthalic anhydride) HOCH₂CHCH₂OH | OH	−COCH₂CHCH₂O−	Cross-linked polyester, paints
Polyesters (Casting resin)	HOCCH=CHCOH HO(CH₂)₄OH	−CCH=CHC−O(CH₂)₂O−	Cross-linked with styrene and benzoyl peroxide, fiberglass boat resin, casting resin
Phenol-formaldehyde (Bakelite)	OH (phenol) CH₂=O	(phenol-formaldehyde network) −CH₂−, CH₂−, CH₂−	Mixed with fillers, molded electrical cases, adhesives, laminates, varnishes

TABLE 8.2 (*Continued*)

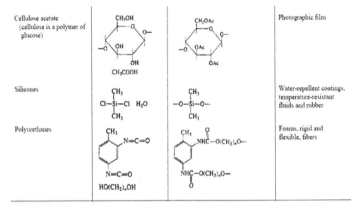

Cellulose acetate (cellulose is a polymer of glucose)			Photographic film
Silicones			Water-repellent coatings, temperature-resistant fluids and rubber
Polyurethanes			Foams, rigid and flexible, fibers

8.3.3 SYNTHETIC POLYMERS

In our everyday lives we are surrounded by a seemingly endless range of synthetic polymers. Synthetic polymers include materials that we customarily call plastics and rubbers. These include such commonplace and inexpensive items as polyethylene grocery sacks, polyethylene terephthalate soda bottles, and nylon backpacks. At the end of the scale are less common and much more expensive polymers that exhibit specialized properties, including Kevlar bullet proof vests, polyacetal gears in office equipment, and high-temperature resistant fluorinated polymer seals in jet engines.

The first completely synthetic polymer (Bakelite) was not made until 1905 and the industry did not really take off until after the Second World War, with developments in methods for polymerizing polyethylene and polypropylene as well as in the production of synthetic fibers. The basic chemistry principles involved in polymer synthesis have not changed much since the beginning of polymer production. Major changes in the last 70 years have occurred in the catalyst field and in process development. Synthetic polymers can be divided into three broad based divisions:

1. *Fibers*: The well known examples are nylon and terylene. They have high strength in one dimension that can be processed into long strands. The range of strength for fibers is truly impressive, from those used for textiles (clothing, carpets; relatively weak), to

commercial monofilament fishing line upto 130km long, to materials that can be woven into bulletproof vests (aramids).

2. *Synthetic rubbers*: These are not produced merely as substitutes for natural rubber. Some of these synthetic rubbers have properties superior to those of the natural product and are thus better suited for specific purposes, for example, neoprene rubber.

3. *Plastics*: These were originally produced as substitutes for wood and metals. Now they are important materials in their own right and tailor-made molecules are synthesized for a particular use, for example, heat resistant polymers which are used in rocketry.

8.3.4 NATURAL POLYMERS

As their name implies, natural polymers occur in nature. Also referred to as biopolymers, they are synthesized in the cells of all organisms. It is interesting to note that two of the most prevalent types, polysaccharides and proteins, each contain diverse compounds with extremely different properties, structures, and uses. For example, the protein in egg white (albumin) serves a much different function (nutrition) from that in silk or wool (structural). Likewise, the properties of starch and cellulose could hardly be more different. Although each is made up of polymers based on condensation of glucose, the final molecular structure differ dramatically. Both sustain life, but in completely different ways.

The utilization of natural polymers for non food uses can be traced back to ancient times. Skin and bone parts of animals, plant fibers, starch, silk, and so on are typical examples of the natural polymers used in different periods of the human history. Modern technologies provide new insights of the synthesis, structures, and properties of the natural polymers.

STARCH POLYMER

Starch (Figure 8.6) consists of two major components: amylase, a mostly linear α-D(1-4)-glucan and amylopectin, an α-D(1-4)-glucan and an α-D(1-6) linkages at the branch point.

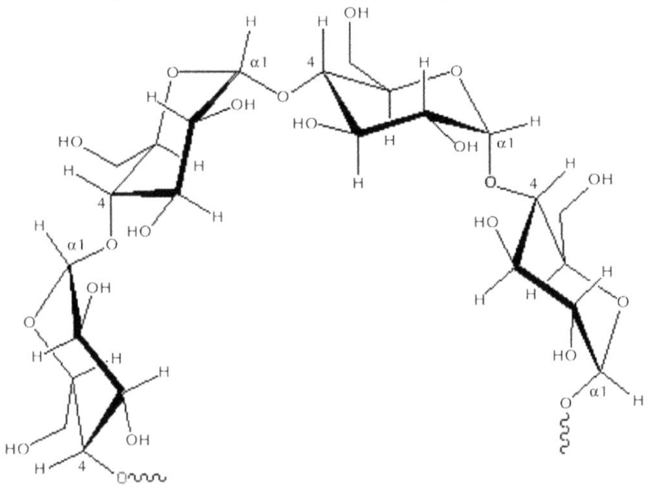

FIGURE 8.6   A short section of starch with four glucose molecules.

Starch is unique among carbohydrates because it occurs naturally as discrete granules. This is because the short branched amylopectin chains are able to form helical structures with crystals. Starch granules exhibit hydrophilic properties and strong inter-molecular association via hydrogen bonding due to the hydroxyl groups on the granule surface. The melting point of native starch is higher than the thermal decomposition temperature: Hence, the poor thermal stability of native starch and the need for conversion to starch based materials with a much improved property profile. Also starch has been intensively studied in order to process it into a thermoplastic in the hope of partially replacing some petrochemical polymers. In its natural form, starch is not meltable and therefore cannot be processed as a thermoplastic. However, starch granules can be thermoplasticized through a gelatinization process.

8.3.5   PROTEINS

The word "protein" is derived from the Greek word prôtos, meaning "primary" or "first rank of importance". Proteins are composed of amino acids and although there are hundreds of different proteins they all consist of linear chains of the same twenty L-α-amino acids which are linked through peptide bonds formed by a condensation reaction. Each amino acid contains an amino group and carboxylic acid group with the general formula:

$$NH_2-CH-R-COOH$$

The R group differs for the various amino acids. The amino acid units are linked together through a peptide bond to form a polypeptide chain. Removing one amino acid or changing it from the protein sequence can be detrimental to its structure and in the same way its biological meaning (Figure 8.7).

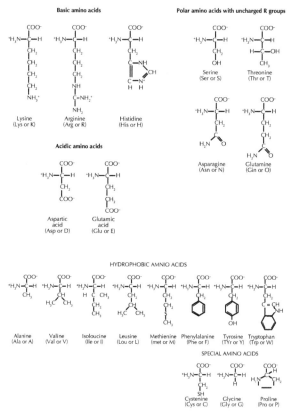

FIGURE 8.7 Molecular structure of 20 essential amino acids that are present in all living organisms. A linear chain of amino acids linked by the amino group of one residue to the carboxyl group of the next makes up a protein.

Proteins can be divided into two main types, based on their overall shape. *Fibrous proteins*, as their name implies, have fiber like structures and are used for structure or support. They are tough and insoluble macromolecules, often having several α-helical chains wound together into rope like bundles. In sharp contrast, the shape of *globular* proteins is spherical or ellipsoidal. They are soluble in water and their solutions have low viscosities.

8.4 PRODUCTION OF NANOFIBERS

Electrospinning [1, 2] is an economical and simple method used in the preparation of polymer fibers. The fibers prepared via this method typically have diameters much smaller than is possible to attain using standard mechanical fiber-spinning technologies [3]. Electrospinning of polymer solutions has gain much attention in the last few years as a cheap and straightforward method to produce nanofibers [4]. Electrospinning differs from the traditional wet or dry fiber spinning in number of ways, of which the most striking differences are the origin of pulling force and the final fiber diameters. The mechanical pulling forces in the traditional industrial fiber spinning processes lead to fibers in the micrometer range and are contrasted in electrospinning by electrical pulling forces that enable the production of nanofibers [5]. Depending on the solution properties, the throughput of single-jet electrospinning systems ranges around 10ml/min. This low fluid throughput may limit the industrial use of electrospinning. A stable cone-jet mode followed by the onset of the characteristic bending instability, which eventually leads to great reduction in the jet diameter; necessitate the low flow rate [6]. When the diameters of polymer fiber materials are shrunk from micrometers (e.g. 10–100mm) to submicrons or nanometers, there appear several amazing characteristics such as very large surface area to volume ratio (this ratio for a nanofiber can be as large as 10^3 times of that of a microfiber), flexibility in surface functionalities, and superior mechanical performance (e.g. stiffness and tensile strength) compared with any other known form of the material. These outstanding properties make the polymer nanofibers to be optimal candidates for many important applications [7]. These include filter media [8], composite materials [9], biomedical applications (tissue engineering scaffolds [10], bandages [11], drug release systems [12]), protective clothing for the military [13], optoelectronic devices and semi-conductive materials [14], and biosensor or chemosensor [15].

A schematic diagram to interpret electrospinning of polymer nanofibers is shown in Figure 8.8. There are basically three components to fulfill the process: a high voltage supplier, a capillary tube with a pipette or needle of small diameter, and a metal collecting screen [16–20].

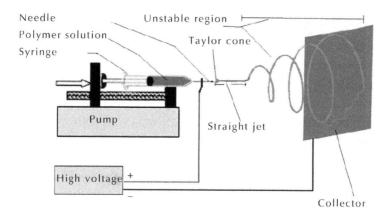

FIGURE 8.8   Electrospinning set up.

8.4.1 PROCESSING CONDITION

APPLIED VOLTAGE

An increase in the electrospinning current generally reflects an increase in the mass flow rate from the capillary tip to the grounded target when all other variables (conductivity, dielectric constant, and flow rate of solution to the capillary tip) are held constant [21, 22].

FEED RATE

The morphological structure can be slightly changed by changing the solution flow rate as shown in Figure 8.9. At the flow rate of 0.3ml/hr, a few of big beads were observed on fibers. The flow rate could affect electrospinning process. A shift in the mass-balance resulted in sustained but unstable jet and fibers with big beads were formed [23].

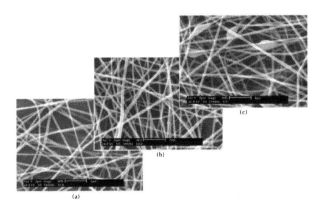

FIGURE 8.9 Effect of flow rate of 7% PVA water solution on fiber morphology (DH = 98%, voltage = 8kV, and tip-target distance = 15cm). Flow rate: (a) 0.1ml/hr; (b) 0.2ml/hr; (c) 0.3ml/hr. Original magnification 10k.

8.5 THEORY

Typically, electrospinning has two stages. In the first, the polymer jet issues from a nozzle and thins steadily and smoothly downstream. In the second stage, the thin thread becomes unstable to a non axisymmetric instability and spirals violently in large loops. For the steady stretching in stage one, Spivak and Dzenis [24] published a simple model that assumes the electric field to be uniform and constant, unaffected by the charges carried by the jet. Hohman et al. [25, 26] developed a slender-body theory for electrospinning that couples jet stretching, charge transport, and the electric field. The model encounters difficulties, however, with the boundary condition at the nozzle.

For stage two, the bending instability has been carefully documented by two groups)Reneker et al. [27, 28], Shin et al. [29]); each has proposed a theory for the instability. Hohman et al. [25] built an electrohydrodynamic instability theory, and predicted that under favorable conditions, a non axisymmetric instability prevails over the familiar Rayleigh instability and a varicose instability due to electric charges. The jet is governed by four steady-state equations representing the conservation of mass and electric charges, the linear momentum balance, and Coulomb's law for the E field [26–29]. Mass conservation requires that

$$\pi R^2 \upsilon = Q \tag{8.1}$$

where, Q is a constant volume flow rate. Charge conservation may be expressed by

$$\pi R^2 KE + 2\pi R \upsilon \sigma = I \tag{8.2}$$

where, E is the z component of the electric field, K is the conductivity of the liquid, and I is the constant current in the jet. The momentum equation is formulated by Figure 8.10.

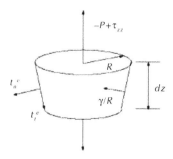

FIGURE 8.10   Momentum balance on a short section of the jet

$$\frac{d}{dz}\left(\pi R^2 \rho \upsilon^2\right) = \pi R^2 \rho g + \frac{d}{dz}\left[\pi R\left(-P + \tau_{zz}\right)\right] + \frac{\gamma}{R}.2\pi RR' + 2\pi R\left(t_t^e - t_n^e R\right), \tag{8.3}$$

where, τ_{zz} is the axial viscous normal stress, p is the pressure, γ is the surface tension, and t_t^e and t_n^e are the tangential and normal tractions on the surface of the jet due to electricity. The prime indicates derivative with respect to z, and R' is the slope of the jet surface. The ambient pressure has been set to zero. The electrostatic tractions are determined by the surface charge density and the electric field:

$$t_n^e = \left\| \frac{\varepsilon}{2}(E_n^2 - E_t^2) \right\| \approx \frac{\sigma^2}{2\varepsilon} - \frac{\varepsilon' - \varepsilon}{2} E^2 \qquad (8.4),$$

$$t_t^e = \sigma E_t \approx \sigma E \qquad (8.5),$$

where ε and $\overline{\varepsilon}$ are dielectric constants of the jet and the ambient air, respectively, E_n and E_t are normal and tangential components of the electric field at the surface, and $\| * \|$ indicates the jump of a quantity across the surface of the jet. We have used the jump conditions for E_n and E_t: $\| \varepsilon E_n \| = \overline{\varepsilon}\,\overline{E} - \varepsilon E_n = \sigma$, $\| E_t \| = \overline{E}_t - E_t = 0$, and assumed that $\varepsilon E_n \ll \overline{\varepsilon}\,\overline{E}$ and $E_t \approx E$. The overbar indicates quantities in the surrounding air. The pressure $p(z)$ is determined by the radial momentum balance, and applying the normal force balance at the jet surface leads to:

$$-p + \tau_{rr} = t_n^e - \frac{\gamma}{R} \qquad (8.6),$$

Inserting Equations (8.4) and (8.6) into Equation (8.3) yields:

$$\rho\upsilon\upsilon' = \rho g + \frac{3}{R^2}\frac{d}{dz}(\eta R^2 \upsilon') + \frac{\gamma R'}{R^2} + \frac{\sigma\sigma'}{\overline{\varepsilon}} + (\varepsilon - \overline{\varepsilon})EE' + \frac{2\sigma E}{R} \qquad (8.7),$$

modeling of the electrospinning process is valuable for the factors perception that cannot be measured experimentally. As mentioned before, although electrospinning gives continuous fibers, mass production and the ability to control nanofibers properties are not obtained yet. On the other hand, during this procedure, the nanofibers gather in a random state on the collector plate; while, in applications such as tissue engineering, well-oriented nanofibers are needed. Mathematical and theoretical modeling of electrospinning process has been focused by many researchers to solve these problems. Recently, several modeling and simulations were introduced to present a better understanding of electrospinning jet mechanics. The development of a relatively simple model of the process was limited by the lack of systematic, fully characterized experimental observations. The governing parameters on electrospining process which are effective on the macroscopic nanofiber properties and commonly investigated by the modeling are solution volumetric flow rate, polymer weight concentration, molecular weight, the applied voltage and the nozzle to ground distance. Polymeric jet ejection is focused for simulation and modeling more than other parts of electrospinning process.

As discussed already, by overcoming surface tension, the polymer jet would be ejected from the nozzle. In the first stage the jet thins until the instability point. In the second stage the jet becomes unstable and undergoes a whipping process. The jet then leaves behind a thin thread of polymer on the collector. According to this fact, two important modeling zones have been introduced so far, which are: a) The zone close to the capillary (jet initiation zone) outlet of the jet and b) The whipping instability zone in which the jet spirals and accelerates towards the collector plate.

It is necessary to study the parameters which influence the nature and diameter of the final fiber to obtain the ability to control morphology. For selected applications, it is desirable to control not only the fiber diameter, but also the internal morphology. An ideal operation is one in which: the nanofibers diameter is be controllable, the surface of the fibers is intact and a single fiber would be collectable. The control of the fiber diameter can be affected by the solution concentration, the electric field strength, the feeding rate at the needle tip and the gap between the needle and the collecting screen.

Concentration of the polymer solution is the most important effective parameters on final nanofibers morphology. Depending on solution concentration during experiments, various impacts for the electric field were observed. Since the effects of solution concentration and electric field strength on mean fiber diameter change at different spinning distances, some interactions and coupling effects are present between the parameters. Figure 8.11 presents the effect of process parameters on the fiber diameter.

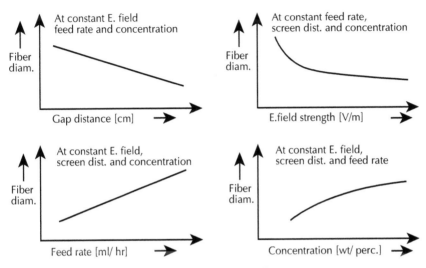

FIGURE 8.11   Effect of process parameters on fiber diameter.

Due to the disadvantages of electrospinning method like low production rate, non oriented nanofiber production; difficulty in diameter prediction and controlling nanofiber morphology, absence of enough information on rheological behavior of polymer solution and difficulty in precise process control which emphasis necessity of modeling. Generally, a suitable theoretical model of the electrospinning process is one which can show a strong-moderate-minor rating effects of these effective parameters on the fiber morphology.

Understanding the effects of viscoelasticity on electrospinning process is important because of its relevance to the creation of fibers made from polymers, glasses, and combinations thereof. In spite of this fact, most of researches in modeling area are limited to study of newtonian fluids, and so do not accurately describe the rheology of viscoelastic materials.

Viscosity plays an important role in determining the rate of thinning of the jet during jet elongation. Researchers have furthermore extended the slender jet model to include viscoelastic effects which will discuss later. The primary role of extensional thickening is to alter the shape of the jet in the intermediate regime of the thinning jet, resulting in sharply thinning cone-jets. Recent experimental studies of model Boger fluids confirm the rapid initial thinning of the cone-jet, for viscoelastic fluids, and suggest the emergence of a new regime where the scaling of diameter with distance along the jet increases from the inverse 1/4 power, typical of Newtonian fluids, towards the inverse 1/2 power for fluids with strong extensional thickening.

Recently, significant progress is made in the development of numerical algorithms for the stable and accurate solution of viscoelastic flow problems, which exists in methods like electrospinning process. There are still some limitations in this section in applying mixed finite element methods to solve viscoelastic flows using constitutive equations of the differential type. Due to the importance of these analyses, some aspects of viscoelastic flow analysis are mentioned briefly in following before discussing about modeling details.

The analysis of viscoelastic flows includes the solution of a coupled set of partial differential equations. The equations commonly depicting the conservation of mass, momentum and energy, and constitutive equations for a number of physical quantities present in the conservation equations such as density, internal energy, heat flux, stress, etc depend on process which will be compeletly discussed later.

The constitutive equation may be transformed into an ordinary differential equation (ODE). For transient problems this can, for instance, be achieved in a natural manner by adopting a Lagrangian formulation. By introducing a selective implicit/explicit treatment of various parts of the equations, a certain separating at each step of the set of equations may be obtained to improve computational efficiency. This suggests the possibility to apply devoted solvers to sub-problems of each fractional time step.

8.5.1 LATTICE BOLTZMANN METHOD (LBM)

An efficient numerical modeling for complex structure web with a reliable prediction needs a complete description of the structure that it is highly difficult. One of these models with agreement by experiments is lattice Boltzmann method Developing lattice Boltzmann method instead of traditional numerical techniques like finite volume, finite difference and finite element for solving large-scale computations and problems involving complex fluids, colloidal suspensions, and moving boundaries is very use-ful.

8.5.2 HYDRODYNAMICS

As nanofibers are made of polymeric solutions forming jets, it is necessary to have a basic knowledge of fluid hydrodynamics. According to the ef-fort of finding a fundamental description of fluid dynamics, the theory of continuity was implemented. The theory describes fluids as small elemen-tary volumes (Figure 8.12), which still are consisted of many elementary particles. Some important equations of fluid hydrodynamics due to this theory are as below:

The equation of continuity:

$$\frac{\partial \rho_m}{\partial t} + div(\rho_m v) = 0 \quad \text{(For incompressible fluids } div(v) = 0 \quad (8.8)$$

The Euler's equation simplified for electrospinning

$$\frac{\partial v}{\partial t} + \frac{1}{\rho_m} \nabla p = 0 \tag{8.9}$$

The equation of capillary pressure:

$$P_c = \frac{\gamma \partial^2 \varsigma}{\partial x^2} \tag{8.10}$$

The equation of surface tension:

$$\Delta P = \gamma \left(\frac{1}{R_X} + \frac{1}{R_Y} \right) \quad R_x \text{ and } R_y \text{ are radii of curvatures (8.11)}$$

The equation of viscosity:

$$\tau_{ij} = \eta \left(\frac{\partial v_i}{\partial x_j} + \frac{\partial v_j}{\partial x_i} \right) \quad \text{(For incompressible fluids, } \tau_{ij} = \text{Stress tensor) (8.12)}$$

$$V = \frac{\eta}{\rho_m} \text{ (Kinematic viscosity) (8.13)}$$

In 1966, Taylor discovered that finite conductivity enables electrical charge to accumulate at the drop interface, permitting a tangential electric stress to be generated. The tangential electric stress drags fluid into motion, and thereby generates hydrodynamic stress at the drop interface. The complex interaction between the electric and hydrodynamic stresses causes either oblate or prolate drop deformation, and in some special cases keeps the drop from deforming.

Scientists used a general treatment of Taylor–Melcher for stable part of electrospinning jets by one dimensional equations for mass, charge, and momentum. In this model, a cylindrical fluid element is used to show electrospinning jet kinematic measurements.

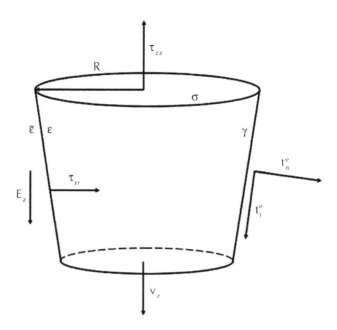

FIGURE 8.12   Scheme of the cylindrical fluid element used in electrohydrodynamic modeling.

In Figure 8.12, the essential parameters are: radius, R, velocity, v_z, electric field, E_z, total path length, L, interfacial tension, γ, interfacial charge, σ, tensile stress, τ, volumetric flow rate, Q, conductivity, K, density, ρ, dielectric constant, $\varepsilon\varepsilon$, and zero-shear rate viscosity, η_0. The most important equations that Feng used are:

$$\tilde{R}^2 \tilde{v}_z = 1 \tag{8.14}$$

$$\tilde{R}^2 \tilde{E}_z + Pe_e \tilde{R}\tilde{v}_z\tilde{\sigma} = 1 \tag{8.15}$$

$$\tilde{v}_z\tilde{v}'_z = \frac{1}{Fr} + \frac{\tilde{T}'}{Re_j \tilde{R}^2} + \frac{1}{We}\frac{\tilde{R}'}{\tilde{R}^2} + \varepsilon(\tilde{\sigma}\tilde{\sigma}' + \beta\tilde{E}_z\tilde{E}'_z + \frac{2\tilde{\sigma}\tilde{E}_z}{\tilde{R}}) \tag{8.16}$$

$$\tilde{E}_z = \tilde{E}_0 - \ln\chi\left[(\tilde{\sigma}\tilde{R})' - \frac{\beta}{2}(\tilde{E}\tilde{R}^2)''\right] \tag{8.17}$$

$$E_0 = \eta_0 v_0 \Big/ R_0 \quad \beta = (\varepsilon/\tilde{\varepsilon}) - 1 \quad \tau = R^2(\tilde{\tau}_{zz} - \tilde{\tau}_{rr}) \tag{8.18}$$

Researchers solved equation under different fluid properties, particularly for non Newtonian fluids with extensional thinning, thickening, and strain hardening, but Researchers developed a simplified understanding of electrospinning jets based on the evolution of the tensile stress due to elongation.

The initialization of instability on the surfaces of liquids is done by applying the external electric field, inducing electric forces on surfaces of liquids. A localized approximation was developed to calculate the bending electric force which acts on an electrified polymer jet, as an important element of the electrospinning process for manufacturing of nanofibers. Using this force, a far reaching analogy between the electrically driven bending instability and the aerodynamically driven instability was established. The description of the wave's instabilities is expressed by equations called dispersion laws. The dependence of wavelength on the surface tension is almost linear and the wavelengths between jets are a little bit smaller for lower depths. Dependency of wavelength on electric field strength is exponential. The dispersion law is identified for four groups of dielectrics liquids with using of Clausius-Mossoti and Onsager's relation (nonpolar liquids with finite and infinite depth and weakly polar liquids with finite and infinite depth). According to these relations relative permittivity is a function of parameters like temperature, square of angular frequency, wave length number and reflective index.

8.5.3 DIMENSIONLESS NON NEWTONIAN FLUID MECHANICS METHOD

The best way for analyzing fluid mechanics problems is to convert parameters to dimensionless form. By using this method, the numbers of governing parameters for given geometry and initial condition reduce. The non dimensionalization of a fluid mechanics problem generally starts with the selection of a characteristic velocity, then because of the flow of non Newtonian fluids, the stress depends non linearly on the flow kinematics, the deformation rate is a main quantity in the analysis of these flows. Next step after determining different parameters is to evaluate characteristic values of the parameters. The non dimensionalization procedure is to employ the selected characteristic quantities to obtain a dimensionless version of the conservation equations and to get to certain numbers like Reynolds number and the Galilei number. The excessive number of governing dimensionless groups posses certain difficulties in the fluid mechanics analysis. Finally by using these results in equation and applying boundary conditions it can be achieved to study different properties.

8.5.4 DETECTION OF X-RAY GENERATED BY ELECTROSPINNING

Electrospinning jets which produces nanofibres from a polymer solution by electrical forces are fine cylindrical electrode creating extremely high electric-field strength in their vicinity at atmospheric conditions. However , this quality of electrospinning is only scarcely investigated, and the interactions of the electric fields generated by them with ambient gases are nearly unknown. Scientists reported on the discovery that electrospinning jets generate X-ray beaming up to energies of 20keV at atmospheric conditions. Researchers investigated on the discovery that electrically charged polymeric jets and highly charged gold-coated nanofibrous layers in contact with ambient atmospheric air generate X-ray beams up to energies of hard X-rays.

The first detection of X-ray produced by nanofiber deposition was observed using radiographic films. The main goal of using these films is to understand Taylor cone creation. The X-ray generation is probably dependent on diameters of the nanofibers affected by the flow rate and viscosity. Thus, it is important to find the ideal concentration (viscosity) of polymeric solutions and flow rate to spin nanofibers as thin as possible. The X-ray radiation can produce black traces on the radiographic film. These black traces were made by outer radiation generated by nanofibers and the radiation has to be in the X-ray region of electromagnetic spectra, because the radiation of lower energy is absorbed by the shield film. Radiographic method of X-ray detection is efficient and sensitive. It is obvious that this method did not tell us anything about its spectrum, but it can clearly show its space distribution. The humidity, temperature, and rheological parameters of polymer can affect on the X-ray intensity generated by nanofibers. The necessity of modeling in electrospinning process and a quick outlook of some important models will be discussed as follows.

8.5.5 MODELING

Using theoretical prediction of the parameters effects on jet radius and morphology can significantly reduce experimental time by identifying the most likely values that will yield specific qualities prior to production. All models start with some assumptions and have shortcomings that need to be addressed. The basic principles for dealing with electrified fluids that Taylor discovered are impossible to account for the most electrical phenomena involving fluids flow under the seemingly reasonable assumptions, which the fluid is either a perfect dielectric or a perfect conductor. The reason is that any perfect dielectric still contains a nonzero free charge density. During steady jetting, a straight part of the jet occurs next to the

Taylor cone in which only axisymmetric motion of the jet is observed. This region of the jet remains stable in time. However, further along its path, the jet can be unstable by non axisymmetric instabilities such as bending and branching, where lateral motion of the jet is observed in the part near the collector.

Branching as the instability of the electrospinning jet can happen quite regularly along the jet if the electrospinning conditions are selected appropriately. Branching is a direct consequence of the presence of surface charges on the jet surface as well as of the externally applied electric field. The bending instability leads to drastic stretching and thinning of polymer jets towards nano-scale in cross-section. Electrospun jets also create perturbations similar to undulations, which can be the source of various secondary instabilities leading to non linear morphologies developing on the jets. The bending instabilities occurring during electrospinning was studied and mathematically modeled by viscoelastic dumbbells connected together. Both electrostatic and fluid dynamic instabilities can contribute to the basic operation of the process.

Different stages of electrospun jets were investigated by different mathematical models by one or three dimensional techniques. Physical models, studying the jet profile, the stability of the jet, and the cone-like surface of the jet, were developed due to significant effects of jet shape on fiber qualities. Droplet deformation, jet initiation, and, in particular, the bending instability which control, to a major extent, fiber sizes and properties are controlled apparently predominantly by charges located within the flight jet. An accurate and predictive tool using a verifiable model which accounts for multiple factors would provide means to run many different scenarios quickly without the cost and time of experimental trial-and error.

LEAKY DIELECTRIC MODEL

The principles for dealing with electrified fluids were summarized by Melcher and Taylor. Their research showed that, it is impossible to explain most of the electrical phenomena involving fluids flow given the hypothesis that the fluid is either a perfect dielectric or a perfect conductor, since both the permittivity and the conductivity affect the flow. An electrified liquid always includes free charge. Although the charge density may be small enough to ignore bulk conduction effects, the charge will accumulate at the interfaces between fluids. The presence of this interfacial charge will result in an additional interfacial stress, especially a tangential stress, which in turn will modify the fluid dynamics. When the conductivity is finite, the leaky dielectric model can be used. Saville indicated that in the solution with weak conduction, the jet carries electric charges only on its surface. The electrohydrodynamic theory proposed by Taylor as the leaky dielectric model is capable of predicting the drop deformation in qualitative agreement with the experimental observations.

Although Taylor's leaky dielectric theory provides a good qualitative description for the deformation of a Newtonian drop in an electric field, the validity of its analytical results is strictly limited to the drop experiencing small deformation in an infinitely extended domain. Comprehensive experiments showed a serious difference in this theoretical prediction. Some investigations were done to solve this defect. For example, the leaky dielectric model was modified by consideration the charge transport to examine electro-kinetic effects.

WHIPPING MODEL

Whipping model is a model for the electrospinning process that mathematically depicts the interaction between the electric field and fluid properties. This model basically works by comparing theory with experimental data. The critical point in this model is determining an accurate treatment for more understanding of physical mechanism of charge transport near the nozzle to have a good prediction which it is not accurately conceivable especially for highly conductive fluids. By giving the shape and charge density of a jet, this model can predict both when and how the jet will become unstable.

Although this model may be applied to predict "terminal" jet diameter, the mathematically derived limiting diameter is equated with experimentally measured fiber diameter.

The diameter depends only on the flow rate, electric current, and surface tension:

$$h_t = c(I/Q)^{-2/3} \gamma^{1/3}$$

(8.19)

where, c is constant.

The significant assumptions in the derivation of this "terminal" diameter contain uniform electric field. There is no phase change and no inelastic stretching of the jet. The experimental research showed that this model can be qualitatively valid for selected polymeric solutions including polyethylene oxide (PEO) and polycaprolactone (PCL) in low concentrations.

The simplified version of the model predicts that

- At the early stages of whipping behavior, the amplitude of the whipping jet h is controlled by charge repulsion and inertia, and h grows exponentially with time in agreement with the earlier linear instability analysis.
- At the next stages of whipping, the inertia effects are damped by viscosity, and the envelope of the whipping jet is controlled by the interplay of charge repulsion and stretching viscosity.
- At the end of the whipping or stretching the surface tension balances the charge repulsion and sets the terminal diameter.

SHAPE EVALUATION MODEL OF ELECTROSPINNING DROPLETS

Perception the drops behavior in an electric field is playing a critical role in practical applications. The practical importance in the processes is electric field-driven flow in which improving the rate of mass or heat transfer between the drops and their surrounding fluid. Numerically, investigations about the shape evolution of small droplets attached to a conducting surface that depends on the electric fields intensity. Based on numerical solution of the Stokes equations for perfectly conducting droplets, different scenarios of droplet shape evolution are distinguished. For this purpose, Maxwell stresses and surface tension were investigated. The advantages of this model are that the non Newtonian effect on the drop dynamics is successfully identified on the basis of electrohydrostatics. In addition, the model shows that the deformation and breakup modes of the non Newtonian drops are distinctively different from the Newtonian cases. The restriction in this model is the electrohydrostatics of the drop phase which should be highly conductive and much less viscous.

NON LINEAR MODEL

During large-strain stretching of the polymer, non linear viscoelasticity models must be used by understanding the role of rheology. For example, a simple two-dimensional model can be used for describing formation of barb electrospun polymer nanowires with a perturbed swollen cross-section and the electric charges "frozen" into the jet surface. This model is integrated numerically using the Kutta–Merson method with the adoptable time step. The result of this modeling is explained theoretically as a result of relatively slow charge relaxation compared with the development of the secondary electrically driven instabilities which deform jet surface locally. When the disparity of the slow charge relaxation compared to the rate of growth of the secondary electrically driven instabilities becomes even more pronounced, the barbs transform in full scale long branches. The competition between charge relaxation and rate of growth of capillary and electrically driven secondary localized perturbations of the jet surface is affected not only by the electric conductivity of polymer solutions but also by their viscoelasticity. Moreover, a non linear theoretical model was able to resemble the main morphological trends recorded in the experiments.

COUPLED FIELD FORCES MATHEMATICAL MODEL FOR
ELECTROSPINNING PROCESS

There is not a theoretical model to describe the electrospinning process under the multi-field forces; therefore, a simple model may be very useful to indicate the contributing factors. Modeling this process can be done in two ways: a) the deterministic approach using classical mechanics like Euler approach and Lagrange approach. b) The probabilistic approach uses E-infinite theory and quantum like properties. Many basic properties are harmonious by adjusting electrospinning parameters such as voltage, flow rate and others, and it can offer in-depth inside into physical understanding of many complex phenomena which cannot be fully explain.

SLENDER-BODY MODEL

One-dimensional models are used for inviscid, incompressible, axisymmetric, annular liquid jets under gravity. In this model, regular perturbations for slender or long jets can be expanded by using integral formulations, Taylor's series expansions, weighted residuals, and variational principles.

Some assumptions that applied from Feng's theory for modeling jets and drops are: The jet radius decreases slowly along z direction while the velocity is uniform in the cross section of the jet so it is lead to the non uniform elongation of jet. According to the parameters can be arranged into three categories: process parameters (Q, I, and E_∞), geometric parameters (R_0 and L) and material parameters (ρ, η_0 (the zero-shear-rate viscosity), $\varepsilon \varepsilon, \bar{\varepsilon}$, K, and γ). The jet can be represented by four steady-state equations: the conservation of mass and electric charges, the linear momentum balance and Coulomb's law for the E field.

Mass conservation can be stated by:

$$\pi R^2 \upsilon = Q \tag{8.20}$$

R: Jet radius

The second equation in this modeling is charge conservation that can be stated by:

$$\pi R^2 K E + 2\pi R \upsilon \sigma = I \tag{8.21}$$

The linear momentum balance is:

$$\rho\upsilon\upsilon' = \rho g + \frac{3}{R^2}\frac{d}{dz}(\eta R^2 \upsilon) + \frac{\gamma R'}{R^2} + \frac{\sigma\sigma'}{\bar{\varepsilon}} + (\varepsilon - \bar{\varepsilon})EE' + \frac{2\sigma E}{R} \qquad (8.22)$$

The Coulomb's law for electric field:

$$E(z) = E_\infty(z) - \ln\chi(\frac{1}{\bar{\varepsilon}}\frac{d(\sigma R)}{dz} - \frac{\beta}{2}\frac{d^2(ER^2)}{dz^2}) \qquad (8.23)$$

L: The length of the gap between the nozzle and deposition point

R_0: The initial jet radius

$$\beta = \varepsilon/\bar{\varepsilon} - 1 \qquad (8.24)$$

$$\chi = L/R_0 \qquad (8.25)$$

By these four equations, the four unknown functions R, v, E and σ are identified.

At first step, the characteristic scales such as R_0 and v_0 are denoted to format dimensionless groups. Inserting these dimensionless groups in four equations discussed above the final dimensionless equations is obtained. The boundary conditions of four equations which became dimensionless can be expressed (see Figure 8.13) as:

$$\text{In } z = 0 \; R(0) = 1 \; \sigma(0) = 0$$

$$\text{In } z = \chi \; E(\chi) = E_\infty$$

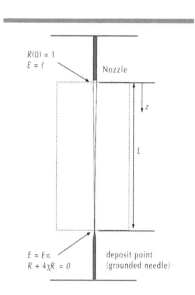

FIGURE 8.13 Boundary conditions.

The first step is to write the ODE's as a system of first order. The basic idea is to introduce new variables, one for each variable in the original problem and one for each of its derivatives up to one less than the highest derivative appearing. In the first step, an initial guess uses for χ and the other parameters would change according to χ. The limitation of slender-body theory is: Avoiding treating physics near the nozzle directly.

VISCOELASTIC FLUIDS MODEL FOR ELECTROSPINNING

When the jet thins, the surface charge density varies, which in turn affects the electric field and the pulling force. Roozemond combined the "leaky dielectric model" and the "slender-body approximation" for modeling electrospinning viscoelastic jets. It must be considered that all variables are uniform on the cross-section of the jet, and vary only along z. The jet could be represented by four steady-state equations: the conservation of mass and electric charges, the linear momentum balance and Coulomb's law for the electric field, with all quantities depending only on the axial position z. The equations can be converted to dimensionless form using some characteristic scales and dimensionless groups like most of the models. These equations could be solved by converting to ODE's forms and using suitable boundary conditions like slender-body model.

MATHEMATICAL MODEL FOR AC-ELECTROSPINNING

Much of the nanofiber research reported so far was about nanofibers which were made from DC potential. In DC-electrospinning, the fiber instability or 'whipping' made it difficult to control the fiber location and the resulting microstructure of electrospun materials. To overcome these limitations, some new technologies were applied in the electrospinning process. The investigations proved that the AC potential resulted in a significant reduction in the amount of fiber 'whipping' and the resulting mats exhibited a higher degree of fiber alignment but are observed to contain more residual solvent. In AC-electrospinning, the jet is inherently unsteady due to the AC potential unlike DC ones, so all thermal, electrical, and hydrodynamics parameters was considered to be effective in the process.

The governing equations for an unsteady flow of an infinite viscous jet pulled from a capillary orifice and accelerated by an AC potential can be expressed as follows:

1. The conservation of mass equation
2. Conservation of charge
3. The Navier–Stokes equation

$$\rho\left(\frac{\partial \upsilon}{\partial t} + v.\nabla v\right) = -\nabla P + \nabla.\tau + F \qquad (8.19)$$

Using these governing equations, final model of AC electrospinning is able to find the relationship between the radius of the jet and the axial distance from nozzle, and a scaling relation between fiber radius and the AC frequency.

ALLOMETRY IN ELECTROSPINNING

Electrospinning applies electrically generated motion to spin fibers; therefore, it is difficult to predict the size of the produced fibers, which depends on the applied voltage in principal. Therefore, the relationship between radius of jet and the axial distance from nozzle was the subject of investigation. It can be described as an allometric equation by using the values of the scaling exponent for the initial steady stage, instability stage, and terminal stage.

The relationship between r and z can be expressed as an allometric equation of the form:

$$r \approx z^b \tag{8.26}$$

where, b is the power exponent. When b = 1 the relationship is isometric and when b ≠ 1 the relationship is allometric.

Assume that the volume flow rate (Q) and the current (I) remains unchanged during the electrospinning procedure, we have the following scaling relations: $Q \sim r_0$, and $I \sim r_0$

Many experiment data shows a power law relationship between electric current and solution flow rate under the condition of fixed voltage:

$$I \approx Q^b \tag{8.27}$$

where, b is the same as the previous equation.

Due to high electrical force acting on the jet, it can be illustrated:

$$\frac{d}{dz}\left(\frac{v^2}{2}\right) = \frac{2\sigma E}{\rho r} \tag{8.28}$$

Equations of mass and charge conservations applied here as mentioned before. From the above Equations, it can be seen that

$$r \approx z^b, \sigma \approx r, E \approx r^{-2}, \frac{du^2}{dz} \approx r^{-2}, \frac{du^2}{dz} \approx r^{-2}, r \approx z^{-\frac{1}{2}} \tag{8.29}$$

The charged jet can be considered as a one-dimensional flow as mentioned. If the conservation equations modified, they would changed as:

$$2\pi r \sigma^{\alpha} v + k \pi r^2 E = I \qquad (8.30)$$

$$r \approx z^{-\alpha/(\alpha+1)} \qquad (8.31)$$

where, α is a surface charge parameter, the value of α depends on the surface charge in the jet.

When $\alpha = 0$ no charge in jet surface, and in $\alpha=1$ use for full surface charge.

MULTIPLE JET MODEL OF ELECTROSPINNING

It was experimentally and numerically exhibited that the jets from multiple nozzles expose higher repulsion by another jets from the neighborhood by Columbic forces than jets spun by a single nozzle process. Researchers achieved upward electrospinning of fibers from multiple jets without the use of nozzles; instead using the spiking effect of a magnetic liquid. For large-scale nanofiber production and the increase in production rate, multi-jet electrospinning systems can be designed to increase both productivity and covering area. The linear Maxwell model and non linear Upper-Convected Maxwell (UCM) model were used to calculate the viscoelasticity. By using these models, the primary and secondary bending instabilities can be calculated. Maxwell model and the non linear UCM model lead to rather close results in the flow dominated by the electric forces. In a multiple-nozzle set up, not only the external applied electric field and self-induced Coulombic interactions influence the jet path, but also mutual-Coulombic interactions between different jets contribute.

MATHEMATICAL MODEL OF THE MAGNETIC ELECTROSPINNING PROCESS

Various forces which induce on polymeric jet make the jet flow whip in a circle and elongate larger and larger. For controlling the instability, magnetic electrospinning is proposed. For describing the magnetic electrospun jet, it can be used Reneker's model. This model may not consider the coupling effects of the thermal field, electric field and magnetic field; therefore, the momentum equation for the motion of the beads is:

$$m\frac{d^2r_i}{dt^2} = F_C + F_E + F_{ve} + F_B + F_q$$

(8.32)

$$m\frac{d^2r_i}{dt^2} F_c + F_{ve} + F_B + F_q$$

Researchers established an electro-magnetic model to explain the bending instability of the charged jet in the electrospinning process. In order to verify the validity of the model, the motion path of the jet was then simulated numerically. The results showed that the magnetic field generated by the electric current in the charged polymer jet may be one of the main reasons which caused the helix motion of the jet during electrospinning.

ELECTROSPINNING NANOPOROUS MATERIALS MODEL

The electrospun nanoporous materials received much attention recently because of the potential for a broad range of applications in widely different areas such as photonic structures, microfluid channels (nanofluidics), catalysis, sensors, medicine, and pharmacy and drug delivery. Due to this importance, some theoretical models were focused on preparing electrospun nanoporous materials with controllable pore sizes and numbers.

Scientists applied the following theoretical model for electrospinning dilation. In this model, the radius of the jet decreases with the increase of the velocity of the incompressible charged jet. The critical jet velocity (maximum velocity before jet changing into unstable form) is calculated as:

$$v_{cr} = Q / \pi\rho r_{cr}^2$$

(8.33)

where, r_{cr} is radius of jet before changing into unstable form.

The velocity can exceed this critical value if a higher voltage is applied. In some cases the radius of the jet reaches the amount of the critical value, and the jet speed increases up to it. In order to keep the conservation of mass equation, the jet dilates by decreasing its density, leading to porosity of the electrospun fibers; this phenomenon is called electrospinning-dilation.

The properties of electrospun nanoporous microspheres can be predicted using the following equation:

$$\pi r_{cr}^{2} \tilde{\rho} \upsilon_{0} = \pi R^{2} \tilde{\rho} \upsilon_{min} = Q \tag{8.34}$$

where, $\tilde{\rho}$ the density of dilated microsphere is, υ_{0} is the velocity of the charged jet at $r = r_{cr}$, R is the maximal radius of the microsphere, υ_{min} is the minimal velocity. Higher voltage means higher value of the jet speed at $r = r_{cr}$, and a more drastic electrospinning-dilation process happens, resulting in a lower density of dilated microsphere, smaller size of the microsphere and smaller pores as well.

8.6   EFFECT OF VOLTAGE AND SPINNING DISTANCE ON MORPHOLOGY AND DIAMETER

Figure 8.14 shows the relationship between mean fiber diameter and electric field with concentration of 15% at spinning distances of 5, 7, and 10cm.

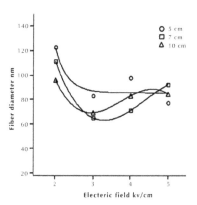

FIGURE 8.14 Momentum balance: The relationship between mean fiber diameter and electric field with concentration of 15% at spinning distances of 5, 7, and 10cm.

The mean fiber diameter obtained at 2kV/cm is larger than other electric fields [31]. Figure 8.15 shows that the concentration apparently has more effect on the fiber diameter than electric field. The multiple regression analysis was carried out to evaluate the contribution of concentration and electric field on the fiber diameter [31–33].

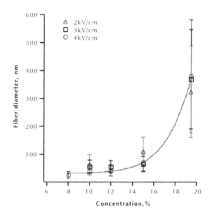

FIGURE 8.15 The relationship between the fiber diameter and the concentration at three electric fields (2, 3, and 4kV/cm).

Researchers used RSM analysis (Response Surface Methodology) to the experimental results to develop a processing window which will produce nanoscale regenerated silk fibers by electrospinning process. The steps in the procedure are described briefly as follows.

1. Identification of variables ζ_1; ζ_2; ζ_3... for response η.
2. Calculation of corresponding coded variables x_1; x_2; x_3... by using the following equation.

$$x_i = \frac{\zeta_i - [\zeta_{Ai} + \zeta_{Bi}]/2}{[\zeta_{Ai} + \zeta_{Bi}]/2} \tag{8.35},$$

where, ζ_{Ai} and ζ_{Bi} refer to the high and low levels of the variables ζ_i; respectively.

3. Determination of the empirical model by multiple regression analysis to generate theoretical responses ($\hat{y}$). The second-order model is widely used in RSM. The general equation for response h of the second-order model is given by:

$$\eta = \beta_0 + \sum_{i=1}^{k} \beta_i x_i + \sum_{i=1}^{k} \beta_{ii} x_i^2 + \sum_{i<j=2}^{k} \beta_{ij} x_i x_j \tag{8.36},$$

where, k is the number of factors, x_i are the coded variables and β are coefficients.

4. Calculation of the coefficients b to fit the experimental data as close as possible.

    When k = 2; the empirical model from the general Equation (8.9) becomes:

$$y = \beta_0 x_1 + \beta_1 x_1 + \beta_2 x_2 + \beta_{11} x_1^2 + \beta_{22} x_2^2 + \beta_{12} x_1 x_2 + \varepsilon \tag{8.37}.$$

8.7   CONCLUDING REMARKS

Nanotechnology has become in recent years a topic of great interest to scientists and engineers, and is now established as prioritized research area in many countries. The reduction of the size to the nano-meter range brings an array of new possibilities in terms of material properties, in particular with respect to achievable surface to volume ratios. Electrospinning of natural fibers is a novel process for producing superfine fibers by forcing a solution through a spinnerette with an electric field. A comprehensive review on this technique has been made in this paper. Based on this review, many challenges exist in the electrospinning process of nanofibers, and number of fundamental questions remains open. The electrospinning technique provides an inexpensive and easy way to produce nanofibers on low basis weight, small fiber diameter, and pore size. It is hoped that this chapter will pave the way toward a better understanding of the application of electrospinning of nanofibers.

Electrospinning is a very simple and versatile method of creating polymer-based high-functional and high-performance nanofibers that can revolutionize the world of structural materials. The process is versatile in that there is a wide range of polymer and biopolymer solutions or melts that can spin. The electrospinning process is a fluid dynamics related problem. In order to control the property, geometry, and mass production of the nanofibers, it is necessary to understand quantitatively how the electrospinning process transforms the fluid solution through a millimeter diameter capillary tube into solid fibers which are four to five orders smaller in diameter. When the applied electrostatic forces overcome the fluid surface tension, the electrified fluid forms a jet out of the capillary tip toward a grounded collecting screen. Although electrospinning gives continuous nanofibers, mass production, and the ability to control nanofibers properties are not yet obtained. Combination of both theoretical and experimental approaches seems to be promising step for better description of electrospinning process.

Applying simple models of the process would be useful in atoning the lack of systematic, fully characterized experimental observations and the theoretical aspects in predicting and controlling effective parameters. The analysis and comparison of model with experiments identify the critical role of the spinning fluid's parameters. The theoretical and quantitative tools developed in different models provide semi- empirical methods for predicting ideal electrospinning process or electrospun fiber properties. In each model, researcher tried to improve the existing models or changed the tools in electrospinning by using another view. Therefore, it was attempted to have a whole view on important models after investigation about basic objects. A real mathematical model, or, more accurately, a real physical model, may initiate a revolution in understanding of dynamic and quantum-like phenomena in the electrospinning process. A new theory is much needed which bridges the gap between Newton's world and the quantum world.

The other important use of nanofibers is production of high performance textile fabrics. According to world statistics, 29% of the global causes of death is heart disease. Once more, the importance of this issue and concern is based on published statistics. Statistics shows that before cardiac arrest, 22% of patients experience chest pain for 120min, 15% suffered asthmatic for 30min, 7% suffer nausea and vomiting for 120min and 5% had been suffered from dizziness. Only 25% of patients had no warning signs before heart attack. The researchers reported that 90% of patients have some sort of signal at least 5min before the symptoms.

Typically, the electrocardiogram (ECG) is the first preliminary test on heart patients. This requires spending lots of time and cost to the patient. In addition, the received information from the patient's ECG shows the situation at test moment. Meanwhile, the heart had a different function in different physical conditions and everyday movement, which is highly encountered to risk of heart stroke. Thus continues access to the heart wave function during daily activities by receiving continuous bioelectric signals is essential.

The ECG system is a sophisticated device for patient vital heart signal monitoring and supervision. An extensive range of human physiological conditions can be deduced from the PQRST waves obtained from an ECG instrument. A typical heart signal wave has been demonstrated in Figure 8.16.

An ECG is a recording of the electrical activity that initiates each hearbeat.

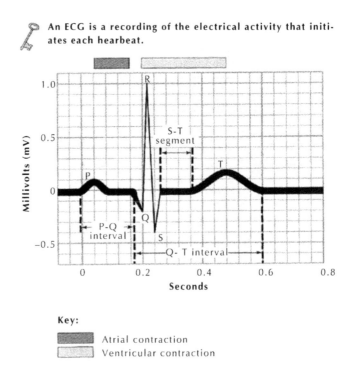

Key:

Atrial contraction
Ventricular contraction

FIGURE 8.16   Typical PQRST heart signal wave.

P wave indicates electrical current in atriums, QRS complex shows electrical current in ventricles and T wave demonstrates heart relaxation period and charging time between two hear beats. The heart muscle tissue is the self-stimulating with spontaneous contractions that convert biochemical energy into kinetic energy, and it provide the needed driving force for the circulation system.

Heart muscle contractions associated with electrical stimulation is called the action potential and bioelectrical signal is triggered. Action potential causes electric shock that passes through cell membranes and controls the amount of contraction in heart muscle strings. Charged ions exchange and displacement from one side to the other side of the cell membrane is carried by ion channels causing action potential. General release in the form of an action potential is shown in Figure 8.17.

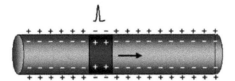

FIGURE 8.17   The release of action potential.

Membranes of heart cells control the number of ions including sodium, potassium, calcium, and chloride ions into and out of the cell. The ions enter and exit take place through a special duct for the same ion. Ion channels transmission is capable for opening and closing due to the changing of cell voltage. In resting heart electrochemical equilibrium between the ions located on both sides of the cell membrane is established in which case the electric potential inside the cell than outside the cell is negative and its value is about 80–90mV. By the time passing, the cell membrane conductivity changes through the opening and closing of the ions passing channels and consequently the potential difference between inside and outside the cell environment changes. In this process the cell initially is de-polarized and again with increasing of the polarity cell potential back to rest. This process is performed on all heart cells, resulting in the contraction of heart muscle and back again to initial form.

Electrical activity of heart cells in the body produces electrical current on the skin resulting in the measurable potential difference. Consequently, the measuring of potential difference on skin surface is an effective way to evaluate the heart performance.

ECG machine combine the received bioelectrical signals from implemented sensors. The attached sensors to certain points of the body are able to detect and measure the changes in the potential difference created in the skin surface. The used sensors in the ECG should be highly electrically conductive because the variation of potential differences in skin surface is very low and is in order of milivolt (mV). In most cases, a combination of silver or silver chloride is used for fabrication of these sensors. Also for better contact with the skin, the conductive gel is used.

Acquired signals using sensors are quite weak and boosting system is extremely necessary to monitor the change in bioelectrical signals. These electrical signals also can be affected by electromagnetic waves, the change in user geographical position and connecting wires. It is therefore, necessary to use filter and or noise depletion circuit. Two types of filters including high pass (HP) and low pass (LP) have been designed for commercial ECG. HP filters adjusts between 0.5 and 1Hz and sort out those signals smaller than threshold limit and reduce the baseline deviation. However, LP filters adjust on 40Hz and remove signals higher than threshold limit. In addition, it can delete the environmental noises between 50 and 60Hz. The above mentioned items besides monitoring system make commercial ECG instruments quite bulky, and miniaturization in recent years has enabled development of wearable versions that collect and process ECG data. Subtle changes in the physiological signals of an individual can be monitored using wireless wearable ECG monitoring system. Telemetry in combination with such systems prepares physicians to take clinical decisions when considerable changes are noticed in the physiological parameters. The conventional physiological monitoring system used in hospitals cannot be used for wearable physiological monitoring applications due to the following reasons.

- The conventional physiological monitoring systems are massive to be used for wearable monitoring.
- The gels used in the electrodes dry out when used over a period of time, which lead to increase in the contact resistance and thereby degrading the signal quality.
- The gels used in the electrodes cause irritations and rashes when used for longer durations.
- There are number of hampering wires from the sensors to the data acquisition system.
- The sensors used in conventional monitoring systems are bulky and are not comfortable to wear for longer durations.

To overcome the above problems associated with the conventional physiological monitoring there is a need to develop sensors for wearable monitoring, integrate them into the fabric of wearer, and continuously monitor the physiological parameters.

Looking at the recent trends in biomedical applications, a major advancement can be noted in non stationary health monitoring devices. This ranges from simple and portable Holter instruments to sophisticated and costly implantable gadgets.

The Holter monitors have been used only to collect data in ambulatory patients. Processing and analysis are then performed offline on recorded data. Systems with multiple sensors have too many wires between the sensors and the monitoring device, which limit the patient's activity and comfort level. Available systems also lack universal connectivity of interfacing to any output display device through common communication ports. There is a requirement of data acquisition (DAQ) circuits with analog to digital converters as the interface between the Instrument and the computer.

The hospitals generally use Wilson Central Terminal arrangement having three electrodes placed at the limbs and connected at the inverting input of the ECG amplifier. However, the Wilson reference can degrade the overall amplifier specifications. Implantable Cardioverter Defibrillator (ICD) is other system, which is very expensive and invasive method to record physiological data. The ICD is used only on high-risk cardiac patients. MOLEC monitor is an embedded real time system that captures, processes, detects, analyzes, and notifies abnormalities in ECG. However, the cost of MOLEC monitor is high. Further, EPI-MEDIC is a twelve-lead ECG system that allows continuous monitoring, but has a large array of electrodes, which makes the system cumbersome. Normal ECG done in clinical setting allows monitoring and recording but gives no analytical results. Available data analysis algorithms are complicated, as they do not implement transparent decision procedure.

The growing health concerns, especially for cardiac disorders reflect on the need of developing a simple and portable ECG system for daily life.

A number of wearable physiological monitoring systems have been put into practical use for health monitoring of the wearer in hospital and real life situations and their performances have been reported. Varying degrees of success have been reported and the percentage failures in the outdoor use are high. The drawbacks associated with the conventional wearable physiological systems are:

- The cables woven into the fabric to interconnect the sensors to the wearable data acquisition hardware pick up interfering noises (e.g., 50Hz power line interference). The wires integrated into fabric act like antennas and can easily pick up the noises from nearby radiating sources.

- The sensors once integrated into the fabric, its location cannot be changed or altered easily.
- The power required for the sensors to operate is to be drawn from the common battery housed in the wearable data acquisition hardware and are routed through wires woven in the fabric.
- Typically these systems consist of a centralized processing unit to digitize process and transmit the data to a remote monitoring system. The processor is loaded heavily to perform multi-channel data acquisition, processing, and transmission of data.
- The cables from the vest interconnecting the sensors can get damaged very easily due to twisting and turning of the cables, while the wearer is performing his routine activity.

To overcome the above issues related to wearable physiological monitoring, the individual sensors integrated into the vest can be housed with electronics and wireless communication system to acquire and transmit the physiological data. The recent advances in the sensors (MEMS and Nanotechnology), low-power microelectronics and miniaturization and wireless networking enable Wireless Sensor Networks (WSN) for human health monitoring. A number of tiny wireless sensors, strategically placed on the human body create a wireless body area network that can monitor various vital signs, providing real-time feedback to the user and medical personnel.

The implications and potentials of these wearable health monitoring technologies are paramount for their abilities to: (1) detect early signs of health deterioration, (2) notify health care providers in critical situations, (3) find correlations between lifestyle and health issues, (4) bring sports conditioning into a new dimension, by providing detailed information about physiological signals under various exercise conditions, and (5) transform health care by providing doctors with multi-sourced, real-time physiological data.

Regarding to these concerns textile material offers new possibilities to bring new advancement to ECG system using intelligent textiles. Previous work on so-called 'intelligent textiles' shows that biomedical monitoring systems can greatly benefit from the integration of electronics in textile materials. This synergy between textiles and electronics mainly originates from the fact that clothing is our most natural interface to the outside world. The higher the level of integration of the sensors and circuits in the clothing, the more unnoticeable the monitoring system becomes to its user and thus the higher the comfort of the user. The elements of wearable ECG system operationally are similar to commercial device but have some basic difference in ECG electrodes and the size of amplifying and filtering circuits.

Various technologies used to design and manufacturing of embedded sensors in smart textile based on their properties such as flexibility, durability, stability during use, and no damage to the skin. Using metallic yarns as a woven matrix or Textrod is the most common method of making textile sensors. In this method, stainless metallic yarns using weaving or knitting techniques can produce seamless texture.

Sensors made by steel yarn textures able to receive and transfer bio-electrical heart signals and transfer heart signals nonetheless these sensors are also have disadvantages. Woven sensors create some problems such as the abrasions on the skin surface from free tail of metallic yarn.

Moreover, the sewn sensors due to low flexibility provide not sufficient comfort for users. The problem can be resolved using circular seamless texture in the context of the fabric sensor, but the location for a specific user is considered by default. For example, a different user needs different cloth according to their size. In another attempt, the flexible sensor is prepared using graphite coated with nickel in the bed of silicon. The problem is connecting of the sensor to the fabric and its connection to the communications system using metallic wire for signal transmission. In accordance to sensor thickness, the user does not feel comfortable. Another way for making wearable sensor is sputter coating. The fabric sensor in this method is coated by copper particle. This sensor due to very low thickness shows good flexibility. Moreover, the fabricated sensor shows very good uniformity and an acceptable quality of received signals. The big problem is the high cost of producing and the low production rate.

In the development of smart textile sensors, another approach is presented which uses electroless metal reduction method for preparation of fabric sensors. This approach can be divided into two groups including electroless plating and electroless printing. The first technique achieves the textile sensor through making a highly conductive fabric. It in turn must be sewn on specified position to touch skin for gathering of bio-electric data. In the latter, however there is no need to sewing. In fact, the sensor as a pattern can be printed on fabric with a very high precision. The major benefit of using this technique is that the measurement devices can be 'seamlessly 'integrated into textile. Hereby, both patient comfort and ease of use in everyday monitoring are enhanced. Also, a single channel ECG monitor has been developed that successfully extracts ECG wave in a simple, wireless and cost effective manner. The amplified and filtered output is then fed to a commercially available cost-effective radio frequency short range Bluetooth transmitter integrated in a pocket PC equipped with software for data analysis. Automatic alarm detection system, can also give early alarm signals even if the patient is unconscious or unaware of cardiac arrhythmias. Simultaneously, a warning message is sent to the doctor at the hospital uses a short message. Also GPRS enable us as to transmit ECG data in a specified time frame to a Clinical Diagnostic Station (CDS). This chapter describes the implementation of and experiences with this new system for wireless monitoring.

KEYWORDS

- **Biomimetics**
- **Electrospinning technology**
- **Nanofiber**
- **Nanotechnology**
- **Polymers**

REFERENCES

1. Wan, Y. Q., Guo, Q., & Pan, N. (2004). *International Journal of Non linear Sciences and Numerical Simulation, 5,* 5.
2. Feng, J. J. (2003) *Journal of Non Newtonian Fluid MechanicsNon,* 116, 55.
3. He, J., Wan, Y., & Yu, J.-Y. (2005). *Polymer,* 46, 2799.
4. Zussman, E., Theron, A., & Yarin, A. L. (2003). *Applied Physics Letters, 82,* 973.
5. Reneker, D. H., Yarin, A. L., Fong, H., & Koombhongse, S. (2000). *Journal of Applied Physics, 87,* 4531.
6. Theron, S. A., Yarin, A. L., & Zussman, E. (2005). *E. Kroll, 46,* 2889.

7. Huang, Z.-M., Zhang, Y.-Z., Kotak, M., & Ramakrishna, S. (2003). *Composites Science and Technology, 63,* 2223.

8. Schreuder-Gibson, H. L., Gibson, P., Senecal, K., Sennett, M., Walker, J., Yeomans, W., et al. (2002). *Journal of Advanced Materials, 34*(3) 44.

9. Ma, Z., Kotaki, M., Inai, R., & Ramakrishna, S. (2005). *Tissue Engineering, 11,* 101.

10. Ma, Z., Kotaki, M., Thomas, Y., Wei, He, & Ramakrishna, S. (2005). *Biomaterials, 26,* 2527.

11. Jin, H. J., Fridrikh, S., Rutledge, G. C., & Kaplan, D. (2002). *Abstracts of Papers American Chemical Society, 224*(1–2), 408.

12. Luu, Y. K., Kim, K., Hsiao, B. S., Chu, B., & Hadjiargyrou, M. (2003) *Journal of Controlled Release, 89,* 341.

13. Schreuder-Gibson, H. L., Gibson, P. W., Senecal, K., Sennett, M., Walker, J., Yeomans, W., et al. (2002). *Journal of Advanced Materials, 34*(3), 44.

14. Senecal, K. J., Samuelson, L., Sennett, M., & Schreuder, G. H. (2001). *US patent application publication* 0045547.

15. Sawicka, K., Goum, P., & Simon, S. (2005). *Sensors and Actuators B, 108,* 585.

16. Fujihara, K., Kotak, M., & Ramakrishn, S. (2005). *Biomaterials, 26,* 4139.

17. Fang, X. & Reneker, D. H. (1997). *Journal of Macromolecular Science Physics, B36,* 169.

18. Taylor, G. I. (1969). *Proceedings of the Royal Society London, Series A, 313,* 453.

19. Kenawy, El-R., Bowlin, G. L., Mansfield, K., Layman, J., Simpsonc, D. G., Sanders, E. H., & Wnek, G. E. (2002). *Journal of Controlled Release, 81,* 57.

20. Fennessey, S. F., & Farris, J. R. (2004). *Polymer, 45,* 4217.

21. Zussman, E., Theron, A., & Yarin, A. L. (2003). *Applied Physics Letters, 82,* 973.

22. Deitzel, J. M., Kleinmeyer, J., Harris, D., & Beck, T. N. (2001). *Polymer, 42,* 261.

23. Zhang, C. H., Yuan, X., Wu, L., Han, Y., & Sheng, J. (2005). *European Polymer Journal, 41,* 423.

24. Spivak, A. F. & Dzenis, Y. A. (1998). *Applied Physics. Letters, 73,* 3067.

25. Hohman, M. M., Shin, M., Rutledge, G., & Brenner, M. P. (2001). *Physics of Fluids, 13,* 2201.

26. Hohman, M. M., Shin, M., Rutledge, G., & Brenner, M. P. (2001). *Applications of Physics of Fluids, 13,* 2221.

27. Reneker, D. H., Yarin, A. L., Fong, H., & Koombhongse, S. (2000). *Journal of Applied Physics, 87,* 4531.

28. Yarin, A. L., Koombhongse, S., & Reneker, D. H. (2001). *Journal of Applied Physics, 89.*

29. Shin, Y. M., Hohman, M. M., Brenner, M. P., & Rutledge, G. C. (2001). *Applied Physics Letters, 78,* 1149.

30. Reneker, D. H., Yarin, A. L., Fong, H., & Koombhongse, S. (2000). *Journal of Applied Physics, 87,* 4531.

31. Sukigara, S., Gandhi, M., Ayutsede, J., Micklus, M., & Ko, F. (2003). *Polymer, 44,* 5727.

32. Sukigara, S., Gandhi, M, Ayutsede, J., Micklus, M., & Ko, F. (2004). *Polymer, 45,* 3708.

33. Park, K. E., Jung, S. Y., Lee, S. J., Min, B-M., & Park, W. H. (2006). *International Journal of Biological Macromolecules, 38,* 165.

CHAPTER 9

NANOMATERIALS: DOS AND DON'TS OF PRODUCTION PROCESS IN LABORATORY

A. K. HAGHI and A. HAMRANG

CONTENTS

9.1   Introduction 197
9.2   Laboratory Set-Up, Production Conditions, and Systematic
       Parameters 199
       9.2.1   Solution Properties 202
       9.2.2   Processing condition 207
9.3   Contact Angle 210
9.4   Case Study I 211
       9.4.1   Materials 211
       9.4.2   Electrospinning 211
       9.4.3   Characterization 211
       9.4.4   Experimental Design 212
9.5   Morphological Analysis of Nanofibers 213
9.6   The Analysis of Variance (ANOVA) 219
9.7   Response Surfaces for AFD 225
       9.7.1   Solution Concentration 225
       9.7.2   Applied Voltage 226
       9.7.3   Tip to Collector Distance 226
       9.7.4   Volume Flow Rate 226

9.8   Response Surfaces for Ca of Electrospun Fiber MAT 228

    9.8.1   Solution Concentration 228

    9.8.2   Applied Voltage.. 228

    9.8.3   Tip to Collector Distance 229

    9.8.4   Volume Flow Rate.. 229

9.9   Determination of Optimal Conditions 230

9.10   Relationship Between AFD and CA of Electrospun Fiber

    MAT 231

9.11   Production of Nanocomposites 232

    9.11.1   Biopolymers 233

    9.11.2   Nanobiocomposites with Chitosan Matrix.................... 236

    9.11.3   Nanotube Composites 239

    9.11.4   Mechanical and Electrical Properties of

            Nanocomposites... 242

9.12   Applications 244

9.13   Chitosan or Carbon Nanotube Nanofluids................................ 245

9.14   Preparation Methods of Chitosan or CNTS Nano-Composites 246

    9.14.1   Solution-casting-evaporation 246

    9.14.2   roperties and Characterization 247

    9.14.3   Crosslinking-casting-evaporation 248

9.15   Case Study II... 253

9.16   Concluding Remarks... 263

Keywords 264

References.. 264

9.1   INTRODUCTION

Electrospinning is an economical and simple method used in the preparation of polymer fibers. The fibers prepared via this method typically have diameters much smaller than what is possible to attain using standard mechanical fiber-spinning technologies [1]. Electrospinning has gained much attention in the last few years as a cheap and straightforward method to produce nanofibers. Electrospinning differs from the traditional wet or dry fiber spinning in a number of ways, of which the most striking differences are the origin of the pulling force and the final fiber diameters. The mechanical pulling forces in the traditional industrial fiber spinning processes lead to fibers in the micrometer range and are contrasted in electrospinning by electrical pulling forces that enable the production of nanofibers. Depending on the solution properties, the throughput of single-jet electrospinning systems ranges around 10ml/min. This low fluid throughput may limit the industrial use of electrospinning. A stable cone-jet mode followed by the onset of the characteristic bending instability, which eventually leads to great reduction in the jet diameter, necessitate the low flow rate [2]. When the diameters of cellulose fiber materials are shrunk from micrometers (e.g., 10–100mm) to submicrons or nanometers, there appear several amazing characteristics such as very large surface area to volume ratio (this ratio for a nanofiber can be as large as 103 times that of a microfiber), flexibility in surface functionalities, and superior mechanical performance (e.g., stiffness and tensile strength) compared with any other known form of the material.

These outstanding properties make the polymer nanofibers to be optimal candidates for many important applications [3]. These include filter media, composite materials, biomedical applications (tissue engineering scaffolds, bandages, and drug release systems), protective clothing for the military, optoelectronic devices and semi-conductive materials, biosensor or chemosensor [4]. Another biomedical application of electrospun fibers that is currently receiving much attention is drug delivery devices. Researchers have monitored the release profile of several different drugs from variety of biodegradable electrospun membranes. Other application for electrospun fibers is porous membranes for filtration devices. Due to the inter-connected network type structure that electrospun fibers form;

they exhibit good tensile properties, low air permeability, and good aerosol protection capabilities. Moreover, by controlling the fiber diameter, electrospun fibers can be produced over a wide range of porosities. Research has also focused on the influence of charging effects of electrospun non woven mats on their filtration efficiency. The filtration properties slightly depended on the surface charge of the membrane, however the fiber diameter was found to have the strongest influence on the aerosol penetration. Electrospun fibers are currently being utilized for several other applications as well. Some of these include areas in nanocomposites. Figure 9.1 compares the dimensions of nanofibers, micro fibers, and ordinary fibers. When the diameters of polymer fiber materials are shrunk from micrometers (e.g., 10–100μm) to sub-microns or nano meters (e.g., 10×10^{-3} -100×10^{-3}μm), there appear several amazing characteristics such as very large surface area to volume ratio, flexibility in surface functionalities, and superior mechanical performance compared with any other known form of material. These outstanding properties make the polymer nanofibers to be optimal candidates for many important applications [4].

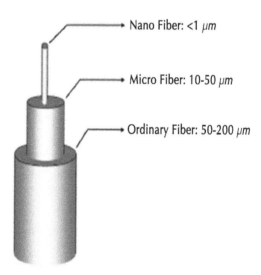

Nano Fiber: <1 μm

Micro Fiber: 10-50 μm

Ordinary Fiber: 50-200 μm

FIGURE 9.1   Classifications of fibers on the basis of fiber diameter.

9.2   LABORATORY SET-UP, PRODUCTION CONDITIONS, AND SYSTEMATIC PARAMETERS

A schematic diagram to interpret electrospinning of nanofibers is shown in Figure. 9.2.

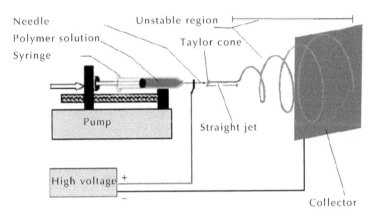

FIGURE 9.2   Electrospinning set up.

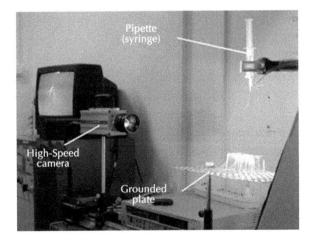

FIGURE 9.3   Electrospinning process.

There are basically three components to fulfill the process: a high voltage supplier, a capillary tube with a pipette or needle of small diameter, and a metal collecting screen. In the electrospinning process (Figure 9.3), a high voltage is used to create an electrically charged jet of polymer solution or melt out of the pipette. Before reaching the collecting screen, the solution jet evaporates or solidifies, and is collected as an interconnected web of small fibers [5]. One electrode is placed into the spinning solution/ melt or needle and the other attached to the collector. In most cases, the collector is simply grounded. The electric field is subjected to the end of the capillary tube that contains the solution fluid held by its surface tension. This induces a charge on the surface of the liquid. Mutual charge repulsion and the contraction of surface charges to the counter electrode cause a force directly opposite to the surface tension [6]. As the intensity of the electric field is increased, the hemispherical surface of the fluid at the tip of the capillary tube elongates to form a conical shape known as the Taylor cone [7]. Further increasing the electric field, a critical value is attained with which the repulsive electrostatic force overcomes the surface tension and the charged jet of fluid is ejected from the tip of the Taylor cone [8]. The jet exhibits bending instabilities due to repulsive forces between the charges carried with the jet. The jet extends through spiraling loops, as the loops increase in diameter the jet grows longer and thinner until it solidifies or collects on the target [9]. A schematic diagram to interpret electrospinning of polymer nanofibers is shown in Figure 9.4 and 9.5. There are basically three components to fulfill the process: a high voltage supplier, a capillary tube with a pipette or needle of small diameter, and a metal collecting screen. In the electrospinning process a high voltage is used to create an electrically charged jet of polymer solution or melt out of the pipette. Before reaching the collecting screen or drum, the solution jet evaporates or solidifies, and is collected as an interconnected web of small fibers. When an electric field is applied between a needle capillary end and a collector, surface charge is induced on a polymer fluid deforming a spherical pendant droplet to a conical shape. As the electric field surpasses a threshold value where electrostatic repulsion force of surface charges overcome surface tension, the charged fluid jet is ejected from the tip of the Taylor cone and the charge density on jet interacts with external field to produce instability.

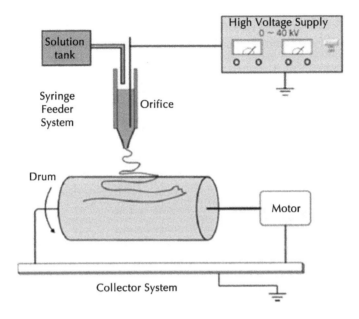

FIGURE 9.4   Schematic diagram of the electrospinning set up by drum collector.

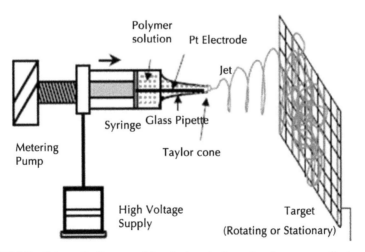

FIGURE 9.5   Schematic diagram of the electrospinning set up by screen collector.

It has been found that morphology such as fiber diameter and its uniformity of the electrospun polymer fibers are dependent on many processing parameters. These parameters can be divided into three groups as shown here:

Solution properties	Viscosity
	Polymer concentration
	Molecular weight of polymer
	Electrical conductivity
	Elasticity
	Surface tension
Processing conditions	Applied voltage
	Distance from needle to collector
	Volume feed rate
	Needle diameter
Ambient conditions	Temperature
	Humidity
	Atmospheric pressure

Under certain condition, not only uniform fibers but also beads-like formed fibers can be produced by electrospinning. Although the parameters of the electrospinning process have been well analyzed in each of polymers, the information has been inadequate enough to support the electrospinning of ultra-fine nanometer scale polymer fibers. A more systematic parametric study is hence required to investigate.

It has been found that morphology such as fiber diameter and its uniformity of the electrospun nanofibers are dependent on many processing parameters. These parameters can be divided into three main groups: a) solution properties, b) processing conditions, and c) ambient conditions. Each of the parameters has been found to affect the morphology of the electrospun fibers.

9.2.1 SOLUTION PROPERTIES

Parameters such as viscosity of solution, solution concentration, molecular weight of solution, electrical conductivity, elasticity, and surface tension, have important effect on morphology of nanofibers.

VISCOSITY

The viscosity range of different nanofiber solution which is spinnable is different. One of the most significant parameters influencing the fiber diameter is the solution viscosity. A higher viscosity results in a large fiber diameter. Figure 9.6 shows the representative images of beads formation in electrospun nanofibers. Beads and beaded fibers are less likely to be formed for the more viscous solutions. The diameter of the beads become bigger and the average distance between beads on fibers longer as the viscosity increases.

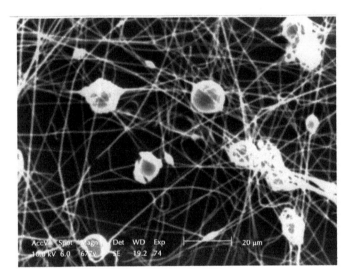

FIGURE 9.6   Electron micrograph of beads formation in electrospun nanofibers.

SOLUTION CONCENTRATION

In electrospinning process, for fiber formation to occur, a minimum solution concentration is required. As the solution concentration increases, a mixture of beads and fibers is obtained (Figure 9.7). The shape of the beads changes from spherical to spindle-like when the solution concentration varies from low to high levels. It should be noted that the fiber diameter increases with increasing solution concentration because of the higher viscosity resistance. Nevertheless, at higher concentration, visco–elastic force which usually resists rapid changes in fiber shape may result in uniform fiber formation. However, it is impossible to electrospin if the solution concentration or the corresponding viscosity become too high due to the difficulty in liquid jet formation.

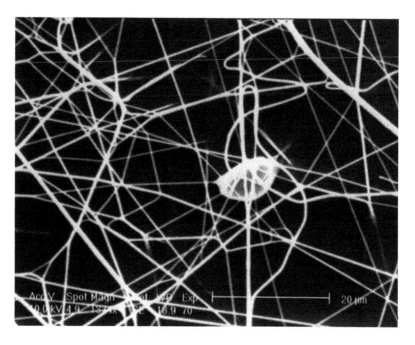

FIGURE 9.7   Electron micrograph of beads and fibers formation in electrospun nanofibers.

MOLECULAR WEIGHT

Molecular weight also has a significant effect on rheological and electrical properties such as viscosity, surface tension, conductivity, and dielectric strength. It has been reported that too low molecular weight solution tend to form beads rather than fibers and high molecular weight nanofiber solution give fibers with larger average diameter (Figure 9.8).

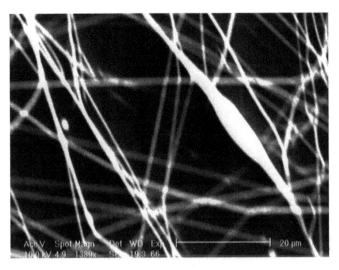

FIGURE 9.8   Electron micrograph of variable diameter formation in electrospun nanofibers.

SURFACE TENSION

The surface tension of a liquid is often defined as the force acting at right angles to any line of unit length on the liquid surface. However, this definition is somewhat misleading, since there is no elastic skin or tangential force as such at the surface of a pure liquid, It is more satisfactory to define surface tension and surface free energy as the work required to increase the area of a surface isothermally and reversibly by unit amount. As a consequence of surface tension, there is a balancing pressure difference across any curved surface, the pressure being greater on the concave side. By reducing surface tension of a nanofiber solution, fibers could be obtained without beads (Figures 9.9 and 9.10). This might be correct in some sense, but should be applied with caution. The surface tension seems more likely to be a function of solvent compositions, but is negligibly dependent on the solution concentration. Different solvents may contribute different surface tensions. However, not necessarily a lower surface tension of a solvent will always be more suitable for electrospinning. Generally, surface tension determines the upper and lower boundaries of electrospinning window if all other variables are held constant. The formation of droplets, bead, and fibers can be driven by the surface tension of solution and lower surface tension of the spinning solution helps electrospinning to occur at lower electric field.

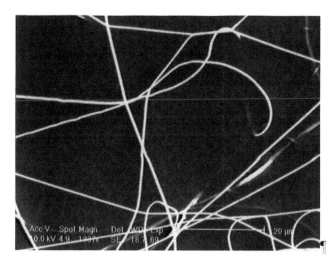

FIGURE 9.9   Electron micrograph of electrospun nanofiber without beads formation.

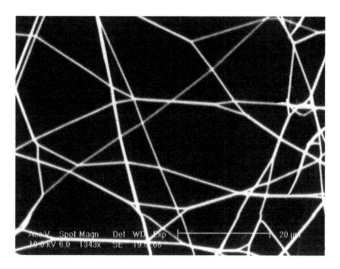

FIGURE 9.10   Electron micrograph of electrospun nanofiber without beads formation.

SOLUTION CONDUCTIVITY

There is a significant drop in the diameter of the electrospun nanofibers when the electrical conductivity of the solution increases. Beads may also be observed due to low conductivity of the solution, which results in insufficient elongation of a jet by electrical force to produce uniform fiber. In general, electrospun nanofibers with the smallest fiber diameter can be obtained with the highest electrical conductivity. This interprets that the drop in the size of fibers is due to the increased electrical conductivity.

9.2.2 PROCESSING CONDITION

APPLIED VOLTAGE

In the case of electrospinning, the electric current due to the ionic conduction of charge in the nanofiber solution is usually assumed small enough to be negligible. The only mechanism of charge transport is the flow of solution from tip to the target. Thus, an increase in the electrospinning current generally reflects an increase in mass flow rate from the capillary tip to the grounded target when all other variables (conductivity, dielectric constant, and flow rate of solution to the capillary tip) are held constant.

With the increase of electrical potential the resulting nanofibers became rougher. It is sometimes reported that a diameter of electrospun fibers does not significantly affected by an applied voltage, this voltage effect is particularly diminished when the solution concentration is low. Applied voltage may affect some factors such as mass of solution fed out from a tip of needle, elongation level of a jet by an electrical force, morphology of a jet (a single or multiple jets), and so on. A balance among these factors may determine a final diameter of electrospun fibers. It should be also noted that beaded fibers may be found to be electrospun with too high level of applied voltage. Although voltage effects show different tendencies, but the voltage generally does not have a significant role in controlling the fiber morphology.

Nevertheless, increasing the applied voltage (i.e., increasing the electric field strength) will increase the electrostatic repulsive force on the fluid jet which favors the thinner fiber formation. On the other hand, the solution will be removed from the capillary tip more quickly as jet is ejected from Taylor cone. This results in the increase of fiber diameter [6-12].

FEED RATE

The morphological structure can be slightly changed by changing the solution flow rate as shown in Figure 9.11. At the flow rate of 0.3ml/hr, a few big beads were observed on the fibers. When the flow rate exceeded a critical value, the delivery rate of solution jet to capillary tip exceeds the rate at which the solution was removed from the tip by the electric forces. This shift in the mass-balance resulted in sustained but unstable jet and fibers with big beads formation.

The solution's electrical conductivity, were found as dominant parameters to control the morphology of electrospun nanofibers [9]. In the case of low-molecular weight liquid, when a high electrical force is applied, formation of droplets can occur. A theory proposed by Rayleigh explained this phenomenon. Evaporation of a droplet takes place and the droplet decreases in size. Therefore, the charge density of its surface is increased. This increase in charge density due to Coulomb repulsion overcomes the surface tension of droplet and causes the droplet to split into smaller droplets. However, in the case of a solution with high-molecular weight liquid, the emerging jet does not break up into droplets, but is stabilized and forms a string of beads connected by a fiber. As the concentration is increased, a string of connected beads is seen, and with further increase there is reduced bead formation until only smooth fibers are formed. And sometimes spindle-like beads can form due to the extension causing by the electrostatic stress. The changing of fiber morphology can probably be attributed to a competition between surface tension and viscosity. As concentration is increased, the viscosity of the solution increases as well. The surface tension attempts to reduce surface area per unit mass, thereby caused the formation of beads or spheres. Visco-elastic forces resisted the formation of beads and allowed for the formation of smooth fibers. Therefore formation of beads at lower solution concentration (low viscosity) occurs where surface tension had a greater influence than the visco-elastic force.

However, bead formation can be reduced and finally eliminated at higher solution concentration, where visco-elastic forces had a greater influence in comparison with surface tension. But when the concentration is too high, high viscosity and rapid evaporation of solvent makes the extension of jet more difficult, thicker and non uniform fibers will be formed [13-20].

Suitable level of processing parameters must be optimized to electrospin solutions into nanofibers with desired morphology and the parameters levels are dependent on properties of solution and solvents used in each of electrospinning process. Understanding of the concept how each of processing parameter affect the morphology of the electrospun nanofibers is essential.

All the parameters can be divided into two main groups; that is, one with parameters which affect the mass of solution fed out from a tip of needle, and the other with parameters which affect an electrical force during electrospinning. Solution concentration, applied voltage and volume feed rate are usually considered to affect the mass.

Increased solution concentration and feed rate tend to bring more mass into the jet. High applied voltage reflects to force to pull a solution out from the needle hence higher applied voltage causes more solution coming out. On the other hand, it should be noted that solution electrical conductivity and applied voltage affect a charge density thus an electrical force, which acts to elongate a jet during electrospinning [21-32].

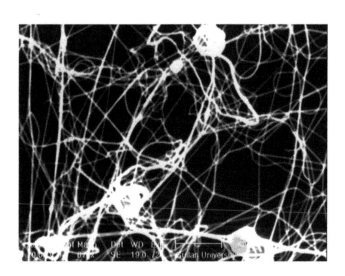

FIGURE 9.11   Electron micrograph of electrospun nanofiber when flow rate exceeded the critical value.

9.3   CONTACT ANGLE

The CA measurements are widely used to characterize the wettability of solid surface. Surface with a water contact angle greater than 150° is usually called superhydrophobic surface. On the other hand, when the contact angle is lower than 5°, it is called superhydrophilic surface. Fabrication of these surfaces has attracted considerable interest for both fundamental research and practical studies [23–25].

In this work, we investigate the effect of four electrospinning parameters (solution concentration, applied voltage, tip to collector distance, and volume flow rate) on the average fiber diameter (AFD) and CA of electrospun polyacrylonitrile (PAN) nanofiber mat. The aim of the present study is to establish quantitative relationship between electrospinning parameters and AFD and CA of electrospun fiber mat by response surface methodology.

9.4 CASE STUDY I

9.4.1 MATERIALS

The PAN powder was purchased from an industrial sector. The average molecular weight (M_w) of PAN was approximately 1,00,000g/mol. The solvent N-N, dimethylformamide (DMF) was obtained from Merck Co. These chemicals were used as received.

9.4.2 ELECTROSPINNING

In our experiment, the PAN powder was dissolved in DMF and gently stirred for 24hr at 50°C. Therefore, homogenous PAN or DMF solution was prepared in different concentration ranged from 10 wt.% to 14 wt.%.

Electrospinning was set up in a horizontal configuration. The electrospinning apparatus consisted of 5ml plastic syringe connected to a syringe pump and a rectangular grounded collector (aluminum sheet). A high voltage power supply (capable to produce 0–40kV) was used to apply a proper potential to the metal needle. It should be noted that all electrospinnings were carried out at room temperature.

9.4.3 CHARACTERIZATION

The morphology of the gold-sputtered electrospun fibers were observed by scanning electron microscope (SEM, Philips XL-30). The average fiber diameter and distribution was determined from selected SEM image by measuring at least 50 random fibers. The wettability of electrospun fiber mat was determined by contact angle measurement. The contact angle measurements were carried out using specially arranged microscope equipped with camera and PCTV vision software as shown in Figure 9.12. The droplet used was distilled water and was 1μl in volume. The contact angle experiments were carried out at room temperature and were repeated five times. All contact angles measured within 20s of placement of the water droplet on the electrospun fiber mat.

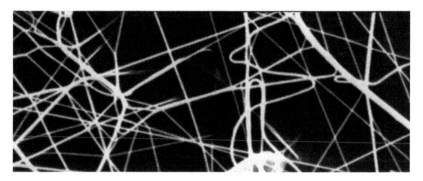

FIGURE 9.12   Schematic of contact angle measurement set up.

9.4.4 EXPERIMENTAL DESIGN

Response surface methodology (RSM) is a combination of mathematical and statistical techniques used to evaluate the relationship between a set of controllable experimental factors and observed results. This optimization process is used in situations where several input variables influence some output variables (responses) of the system. The main goal of RSM is to optimize the response, which is influenced by several independent variables, with minimum number of experiments. Central composite design (CCD) is the most common type of second-order designs that used in RSM and is appropriate for fitting a quadratic surface [26, 27].

In the present study, CCD was employed to establish relationships between four electrospinning parameters and two responses including the AFD and the CA of electrospun fiber mat. The experiment was performed for atleast three levels of each factor to fit a quadratic model. Based on preliminary experiments, polymer solution concentration (X_1), applied voltage (X_2), tip to collector distance (X_3), and volume flow rate (X_4) were determined as critical factors with significance effect on AFD and CA of electrospun fiber mat. These factors were four independent variables and chosen equally spaced, while AFD and CA of electrospun fiber mat were dependent variables (responses). The values of -1, 0, and 1 are coded variables corresponding to low, intermediate, and high levels of each factor respectively. The experimental parameters and their levels for four independent variables are shown in Table 9.1.

TABLE 9.1 Design of experiment (factors and levels)

Factor	Variable	Unit	Factor level		
			−1	0	1
X_1	Solution concentration	(wt.%)	10	12	14
X_2	Applied voltage	(kV)	14	18	22
X_3	Tip to collector distance	(cm)	10	15	20
X_4	Volume flow rate	(ml/hr)	2	2.5	3

The following quadratic model, which also includes the linear model, was fitted to the data.

$$Y = \beta_0 + \sum_{i=1}^{k} \beta_i.x_i + \sum_{i=1}^{k} \beta_{ii}.x_i^2 + \sum \sum_{i<j=2}^{k} \beta_{ij}.x_i x_j + \varepsilon \tag{9.1}$$

where, x_i is the predicted response, x_i and x_j are coded variables, β_0 is constant coefficient, β_i is the linear coefficient, β_{ii} is the quadratic coefficient, β_{ij} is the second-order interaction coefficient, k is the number of factors, and ε is the approximation error [26, 27].

The experimental data were analyzed using Design-Expert software including analysis of variance (ANOVA). The values of coefficients for parameters (β_i, β_i, β_{ii}, and β_{ij}) in Equation 9.1, p-values, the determination coefficient (R^2), and adjusted determination coefficient (R^2_{adj}) were calculated by regression analysis.

9.5 MORPHOLOGICAL ANALYSIS OF NANOFIBERS

The PAN solution in DMF were electrospun under different conditions, including various PAN solution concentrations, applied voltages, volume flow rates, and tip to collector distances, to study the effect of electrospinning parameters on the morphology and diameter of electrospun nanofibers.

Figure 9.13 shows the SEM images and fiber diameter distributions of electrospun fibers in different solution concentration as one of the most effective parameters to control the fiber morphology.

As observed in Figure 9.13, the average fiber diameter increased with increasing concentration. It was suggested that the higher solution concentration would have more polymer chain entanglements and less chain mobility. This causes the hard jet extension and disruption during electrospinning process and producing thicker fibers.

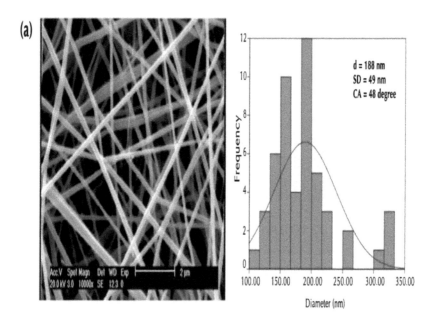

FIGURE 9.13   (*Continued*)

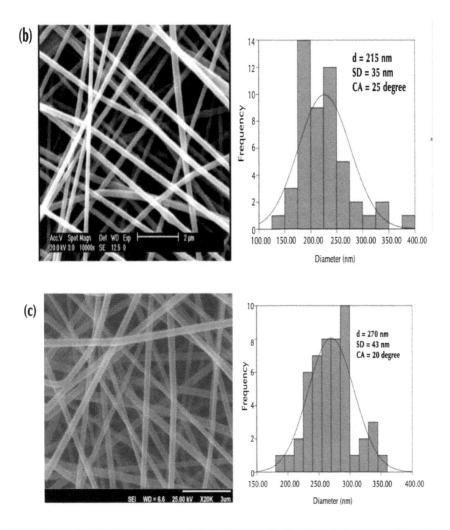

FIGURE 9.13   The SEM images and fiber diameter distributions of electrospun fibers in solution concentration of (a) 10 wt.%, (b) 12 wt.%, and (c) 14 wt.%.

The SEM image and corresponding fiber diameter distribution of electrospun nanofiber in different applied voltage are shown in Figure 9.14. It is obvious that increasing the applied voltage causes an increase followed by a decrease in electrospun fiber diameter. As demonstrated by previous researchers [17, 30], increasing the applied voltage may decrease, increase or may not change the fiber diameter. In one hand, increasing the applied voltage will increase the electric field strength and higher electrostatic repulsive force on the jet, favoring the thinner fiber formation. On the other hand, more surface charge will introduce on the jet and the solution will be removed more quickly from the tip of needle. As a result, the average fiber diameter will be increased [29–30].

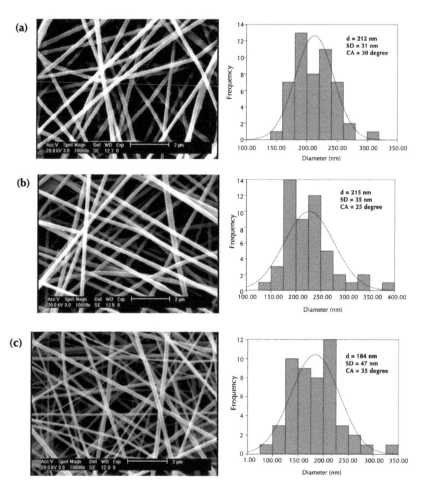

FIGURE 9.14   The SEM images and fiber diameter distributions of electrospun fibers in applied voltage of (a) 14kV, (b) 18kV, and (c) 22kV.

Figure 9.15 represents the SEM image and fiber diameter distribution of electrospun nanofiber in different spinning distance. It can be seen that the average fiber diameter decreased with increasing tip to collector distance. Because the longer spinning distance could give more time for the solvent to evaporate, increasing the spinning distance will decrease fiber diameter [30, 31].

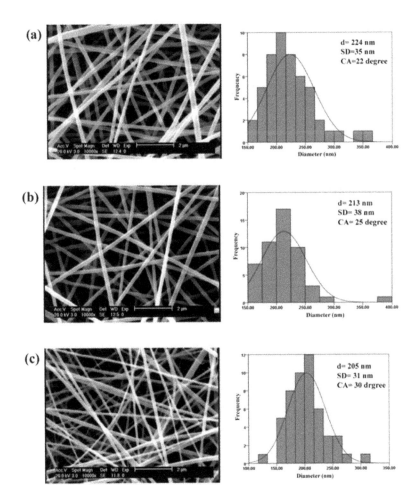

FIGURE 9.15   The SEM images and fiber diameter distributions of electrospun fibers in tip to collector distance of (a) 10cm, (b) 15cm, and (c) 20cm.

The SEM image and fiber diameter distribution of electrospun nanofiber in different volume flow rate are illustrated in Figure 9.16. It is clear that increasing the volume flow rate cause an increase in average fiber diameter. Ideally, the volume flow rate must be compatible with the amount of solution removed from tip of needle. At low volume flow rates, solvent would have sufficient time to evaporate and thinner fibers were produced, but at high volume flow rate, excess amount of solution fed to the tip of needle and thicker fibers result [28–31].

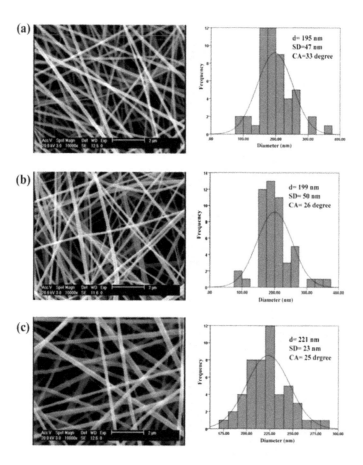

FIGURE 9.16   The SEM images and fiber diameter distributions of electrospun fibers in volume flow rate of (a) 2ml/hr, (b) 2.5ml/hr, and (c) 3ml/hr.

9.6   THE ANALYSIS OF VARIANCE (ANOVA)

All 30 experimental runs of CCD were performed as described in Table 9.2. A significance level of 5% was selected; that is, statistical conclusions may be assessed with 95% confidence. In this significance level, the factor has significant impact on response if the p-value is less than 0.05, and when p-value is greater than 0.05, it is concluded that the factor has no significant effect on response.

TABLE 9.2   The actual design of experiments and responses for AFD and CA

No.	Electrospinning parameters				Responses	
	x_1 Concentration	x_2 Voltage	x_3 Distance	x_4 Flow rate	AFD (nm)	CA (°)
1	10	14	10	2	206 ± 33	44 ± 6
2	10	22	10	2	187 ± 50	54 ± 7
3	10	14	20	2	162 ± 25	61 ± 6
4	10	22	20	2	164 ± 51	65 ± 4
5	10	14	10	3	225 ± 41	38 ± 5
6	10	22	10	3	196 ± 53	49 ± 4
7	10	14	20	3	181 ± 43	51 ± 5
8	10	22	20	3	170 ± 50	56 ± 5
9	10	18	15	2.5	188 ± 49	48 ± 3
10	12	14	15	2.5	210 ± 31	30 ± 3
11	12	22	15	2.5	184 ± 47	35 ± 5
12	12	18	10	2.5	214 ± 38	22 ± 3
13	12	18	20	2.5	205 ± 31	30 ± 4
14	12	18	15	2	195 ± 47	33 ± 4
15	12	18	15	3	221 ± 23	25 ± 3
16	12	18	15	2.5	199 ± 50	26 ± 4
17	12	18	15	2.5	205 ± 31	29 ± 3

TABLE 9.2   (*Continued*)

18	12	18	15	2.5	225 ± 38	28 ± 5
19	12	18	15	2.5	221 ± 23	25 ± 4
20	12	18	15	2.5	215 ± 35	24 ± 3
21	12	18	15	2.5	218 ± 30	21 ± 3
22	14	14	10	2	255 ± 38	31 ± 4
23	14	22	10	2	213 ± 37	35 ± 5
24	14	14	20	2	240 ± 33	33 ± 6
25	14	22	20	2	200 ± 30	37 ± 4
26	14	14	10	3	303 ± 36	19 ± 3
27	14	22	10	3	256 ± 40	28 ± 3
28	14	14	20	3	283 ± 48	39 ± 5
29	14	22	20	3	220 ± 41	36 ± 4
30	14	18	15	2.5	270 ± 43	20 ± 3

The results of analysis of variance (ANOVA) for average fiber diameter and contact angle of electrospun fiber mat are shown in Tables 9.3 and 9.4 respectively. Equations 9.2 and 9.3 are the calculated regression equation.

TABLE 9.3   Analysis of variance for AFD

p-value	F-value	Sum of squares	Source
<0.0001	**28.67**	**31004.72**	**Model**
<0.0001	226.34	17484.50	X_1
<0.0001	54.39	4201.39	X_2
<0.0001	38.04	2938.89	X_3
<0.0001	39.04	3016.06	X_4
0.0016	14.75	1139.06	$X_1 X_2$
0.1524	2.27	175.56	$X_1 X_3$

TABLE 9.3   (*Continued*)

0.0116	8.25	637.56	$X_1 X_4$
0.4879	0.51	39.06	$X_2 X_3$
0.1675	2.10	162.56	$X_2 X_4$
0.3918	0.78	60.06	$X_3 X_4$
0.0032	12.24	945.71	X_1^2
0.0322	5.58	430.80	X_2^2
0.9433	0.005	0.40	X_3^2
0.7334	0.12	9.30	X_4^2
Not significant	0.8	711.41	Lack of fit

$$R^2 = 0.9640; \quad R^2_{adj} = 0.9303.$$

TABLE 9.4   Analysis of variance for CA of electrospun fiber mat

p-value	F-value	Sum of squares	Source
<0.0001	32.70	4175.07	Model
<0.0001	193.01	1760.22	X_1
0.0082	9.27	84.50	X_2
<0.0001	37.06	338.00	X_3
0.0051	10.75	98.00	X_4
0.0116	4.63	42.25	$X_1 X_2$
0.0116	4.63	42.25	$X_1 X_3$
0.0116	4.63	42.25	$X_1 X_4$

TABLE 9.4   (*Continued*)

0.2646	1.34	12.25	X_2X_3
0.4207	0.69	6.25	X_2X_4
0.6266	0.25	2.25	X_3X_4
0.0008	17.75	161.84	X_1^2
0.0039	11.65	106.24	X_2^2
0.9597	0.003	0.024	X_3^2
0.1426	2.40	21.84	X_4^2
Not significant	1.15	95.30	Lack of fit

$$R^2 = 0.9683; \ R^2_{adj} = 0.9387$$

$$AFD = 212.11 + 31.17X_1 - 15.28X_2 - 12.78X_3 + 12.94X_4$$
$$- 8.44X_1X_2 + 3.31X_1X_3 + 6.31X_1X_4 + 1.56X_2X_3 - 3.19X_2X_4 - 1.94X_3X_4$$
$$+ 19.11X_1^2 - 12.89X_2^2 - 0.39X_3^2 - 1.89X_4^2$$

$$(9.2)$$

$$CA = 25.80 - 9.89X_1 + 2.17X_2 + 4.33X_3 - 2.33X_4$$
$$- 1.63X_1X_2 - 1.63X_1X_3 + 1.63X_1X_4 - 0.88X_2X_3 - 0.63X_2X_4 + 0.37X_3X_4$$
$$+ 7.90X_1^2 + 6.40X_2^2 - 0.096X_3^2 + 2.90X_4^2$$

$$(9.3)$$

From the p-values presented in Table 9.3 and 9.4, it can be concluded that the p-values of terms 23X, 24X, 32XX, 31XX, and X_3X_4 in the model of AFD and X_3^2, X_4^2, X_2X_3, X_2X_4 and X_3X_4 in the model of CA is greater than the significance level of 0.05, therefore they have no significant effect on corresponding response. Since the above terms had no significant effect on AFD and CA of electrospun fiber mat, these terms were removed and fitted the equations by regression analysis again. The fitted equations in coded unit are given in Equations 9.4 and 9.5.

$$AFD = 211.89 + 31.17X_1 - 15.28X_2 - 12.78X_3 + 12.94X_4$$
$$- 8.44X_1X_2 + 6.31X_1X_4 \qquad (9.4)$$
$$+ 18.15X_1^2 - 13.85X_2^2$$

$$CA = 26.07 - 9.89X_1 + 2.17X_2 + 4.33X_3 - 2.33X_4$$
$$- 1.63X_1X_2 - 1.63X_1X_3 + 1.63X_1X_4 \qquad (9.5)$$
$$+ 9.08X_1^2 + 7.58X_2^2$$

Now all the p-values are less than the significance level of 0.05.

The predicted versus actual plots for AFD and CA of electrospun fiber mat are shown in Figures 9.17 and 9.18 respectively. Actual values are the measured response data for a particular run and the predicted values evaluated from the model. These plots have determination coefficient (R^2) of 0.9640 and 0.9683 for AFD and CA respectively. It can be observed that experimental values are in good agreement with the predicted values.

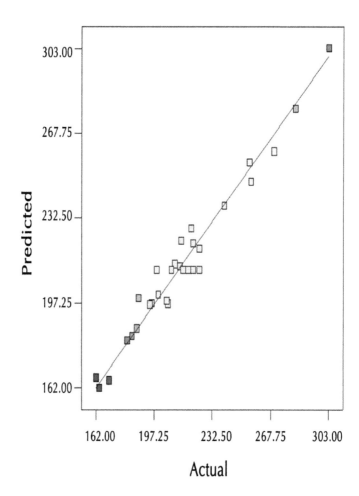

FIGURE 9.17   The predicted versus actual plot for AFD of electrospun fiber mat.

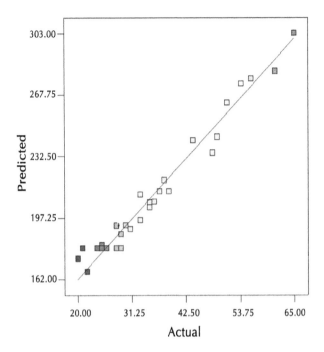

FIGURE 9.18   The predicted versus actual plot for CA of electrospun fiber mat.

9.7   RESPONSE SURFACES FOR AFD

9.7.1   SOLUTION CONCENTRATION

Generally, a minimum solution concentration is required to obtain uniform fibers from electrospinning. Below this concentration, polymer chain entanglements are insufficient and a mixture of beads and fibers is obtained. As the solution concentration increases, the shape of the beads changes from spherical to spindle-like [17].

In the present study, AFD increased with solution concentration as shown in Figure 9.19a, b, c that is in agreement with previous observations [28–30]. Figure 9.19a shows the effect of changing solution concentration and applied voltage at fixed spinning distance and volume flow rate.

It can be seen that AFD increases with increase in solution concentration at any given voltage. As shown in Figure 9.19b, no interaction was observed between solution concentration and spinning distance. This means that the function of solution concentration was independent from spinning distance for AFD. The effect of solution concentration on AFD was influenced by volume flow rate (Figure 9.19c) and this agrees the presence of the term X_1X_4 in the model of AFD.

9.7.2 APPLIED VOLTAGE

Figure 9.19a, d, e show the effect of applied voltage on AFD. In this work, AFD suffer an increase followed by a decrease with increasing the applied voltage. The surface plot in Figure 9.19a indicated that there was a considerable interaction between applied voltage and solution concentration and this is in agreement with the presence of the term X_1X_2 in the model of AFD. In the present study, applied voltage influenced AFD regardless of spinning distance and volume flow rate as shown in Figure 9.19d, e. The absence of X_2X_3 and X_2X_4 in the model of AFD proves this observation.

9.7.3 TIP TO COLLECTOR DISTANCE

The tip to collector distance was found to be another important processing parameter as it influences the solvent evaporating rate and deposition time as well as electrostatic field strength. Figure 9.19b, d, f represents the decrease in AFD with spinning distance. It was founded that spinning distance influences AFD independent from solution concentration, applied voltage, and volume flow rate. This agrees the absence of X_1X_3, X_2X_3 and X_3X_4 in the model of AFD.

9.7.4 VOLUME FLOW RATE

The effect of volume flow rate on AFD is shown in Figure 9.19c, e, f and indicated the increase in AFD with volume flow rate. Figure 9.19c shows the surface plot of interaction between volume flow rate and solution concentration. It can be seen that at fixed applied voltage and spinning distance, the increase in volume flow rate and solution concentration results in higher AFD. As depicted in Figure 9.19e, f, the effect of volume flow rate on AFD was independent from applied voltage and spinning distance. This observation confirms the absence of X_2X_4 and X_3X_4 in the model of AFD.

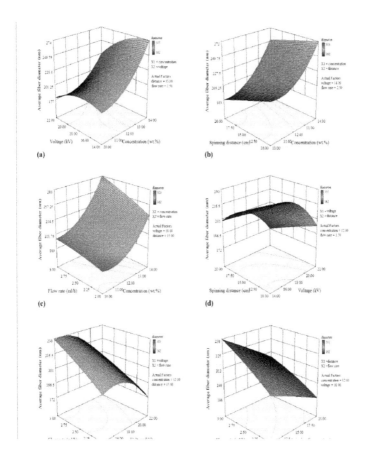

FIGURE 9.19   Response surfaces plot for average fiber diameter showing the effect of: (a) solution concentration and applied voltage, (b) solution concentration and spinning distance, (c) solution concentration and volume flow rate, (d) applied voltage and spinning distance, (e) applied voltage and flow rate, and (f) spinning distance and volume flow rate.

9.8   RESPONSE SURFACES FOR CA OF ELECTROSPUN FIBER MAT

9.8.1   SOLUTION CONCENTRATION

Figure 9.20a, b and c show the effect of solution concentration on CA of electrospun fiber mat. In this work, the CA of electrospun fiber mat decreases with increasing the solution concentration.

Figure 9.20a shows the surface plot of interaction between solution concentration and applied voltage. It is obvious that at fixed spinning distance and volume flow rate, the increase in applied voltage and decrease in solution concentration result the higher CA. As shown in Figure 9.10b, there was a considerable interaction between solution concentration and spinning distance, and this is in agreement with the presence of the term X_1X_3 in the model of CA. The surface plot in Figure 9.20c shows the interaction between solution concentration and volume flow rate at fixed applied voltage and spinning distance. It can be seen that at any given flow rate, CA of electrospun fiber mat will increase as solution concentration decreases.

9.8.2   APPLIED VOLTAGE

The effect of applied voltage on CA of electrospun fiber mat is shown in Figure 9.20a, d, e. It can be seen that the CA suffer a decrease followed by an increase with increasing the applied voltage. As depicted in Figure 9.20a, the impact of applied voltage on CA of electrospun fiber mat will change at different solution concentration. Figure 9.20d shows that there was no combined effect between applied voltage and spinning distance. Also no interaction was observed between applied voltage and volume flow rate (Figure 9.20e). Therefore, applied voltage had interaction with solution concentration which had been confirmed by the existence of term X_1X_2 in the model of CA.

9.8.3   TIP TO COLLECTOR DISTANCE

The impact of spinning distance on CA of electrospun fiber mat is illustrated in Figure 9.20b, d, f. Increasing the spinning distance causes the
CA of electrospun fiber mat to increase. As demonstrated in Figure 9.20b,
low solution concentration cause the increase in CA of electrospun fiber
mat at large spinning distance. Spinning distance affected CA of electrospun fiber mat regardless of applied voltage and volume flow rate (Figure
9.20d, f) as could be concluded from the model of CA. This means that no
interaction exists between these variables.

9.8.4   VOLUME FLOW RATE

The surface plot in Figure 9.20c, e, f represented the effect of volume
flow rate on CA of electrospun fiber mat. Figure 9.20c shows the interaction between volume flow rate and solution concentration. As illustrated in
Figure 9.20 e, f, the effect of volume flow rate on CA of electrospun fiber
mat was independent from applied voltage and spinning distance.

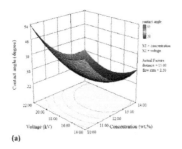

(a)

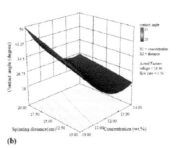

(b)

FIGURE 9.20   *(Continued)*

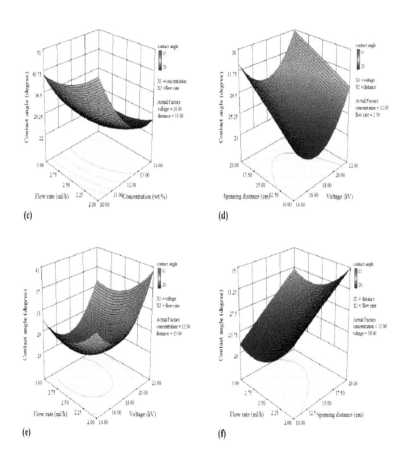

FIGURE 9.20 Response surfaces plot for CA of electrospun fiber mat showing the effect of: (a) solution concentration and applied voltage, (b) solution concentration and spinning distance, (c) solution concentration and volume flow rate, (d) applied voltage and spinning distance, (e) applied voltage and flow rate, and (f) spinning distance and volume flow rate.

9.9   DETERMINATION OF OPTIMAL CONDITIONS

The optimal conditions were established by desirability. Independent variables namely solution concentration, applied voltage, spinning distance, and volume flow rate were set in range and dependent variable (CA) was fixed at minimum. The optimal conditions in the tested range for minimum CA of electrospun fiber mat are shown in Table 9.5.

TABLE 9.5   Optimum values of the process parameters for minimum CA of electrospun fiber mat

Parameter	Optimum value
Solution concentration (wt.%)	13.2
Applied voltage (kV)	16.5
Spinning distance (cm)	10.6
Volume flow rate (ml/hr)	2.5

9.10   RELATIONSHIP BETWEEN AFD AND CA OF ELECTROSPUN FIBER MAT

The wettability of surface controlled by both surface chemistry and surface roughness. The morphology and structure of electrospun fiber mat, such as the nanoscale fibers and interfibrillar distance, increases the surface roughness as well as the fraction of contact area of droplet with air trapped between fibers.

It is proved that the contact angle can provide valuable information about surface roughness. Figure 9.21 shows the variation of CA with AFD. The contact angle is observed to decrease with the increase in average fiber diameter, which is in good agreement with other report [32]. It can be concluded that the thinner fibers, due to their high surface roughness, have higher contact angle than the thicker fibers.

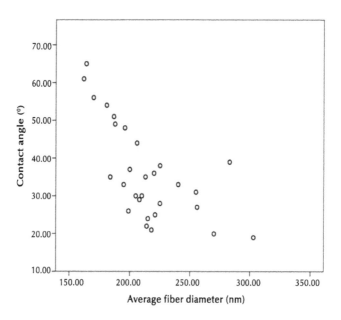

FIGURE 9.21   Variation of contact angle with AFD.

9.11   PRODUCTION OF NANOCOMPOSITES

Recently, terms like "nanobiocomposites" or "biopolymer nanocomposites" are most frequently observed in environmentally friendly research studies. The synthetic polymers have been widely used in a various application of nanocomposites. However, they become a major source of waste after use, due to their poor biodegradability. On the other hand, most of the synthetic polymers show no biocompatibility in *vivo* and *vitro* environments. Hence, scientists were interested in biopolymers as biodegradable materials, so that several groups of natural biopolymers, such as polysaccharides, proteins, and nucleic acids, came to be used in various applications. Nevertheless, the use of these materials has been limited due to their relatively poor mechanical properties. Therefore, research efforts have been made to improve the properties of biopolymers as a matrix by means of reinforcement techniques.

Chitosan (CHT) is a polysaccharide biopolymer widely used as a matrix in nanobiocomposites due to its high biocompatibility and biodegradibility. Therefore, numerous studies have focused on improving the physical properties of biopolymer nanocomposites by using the fundamental behavior of carbon nanotubes.

The present study summarizes the recent advances in the production of carbon nanotubes or chitosan nanocomposites by several methods, special stress being laid on the preparation of CNTs or CHT nanofiber composites by electrospinning method.

9.11.1   BIOPOLYMERS

Biomaterials have been defined as materials biocompatible with the living systems. Biocompatibility implies a chemical, physical (surface morphology) and biological suitability of an implant surface to the host tissues. Over the last 30 years, researchers reviewed various biomaterials and their applications, as well as the applications of biopolymers and their biocomposites for medical applications. These materials can be classified into natural and synthetic biopolymers. Synthetic biopolymers are cheap and have high mechanical properties. The low biocompatibility of synthetic biopolymers compared with that of natural biopolymers, such as polysaccharides, lipids, and proteins, oriented the attention towards natural biopolymers. On the other hand, the natural biopolymers usually have weak mechanical properties. Therefore, many efforts have been done for improving them by blending with some filler.

Among natural biopolymers, polysaccharides seem to be the most promising materials for various biomedical fields. They have various resources including animal origin, plant origin, algal origin, and microbial origin. Among polysaccharides, chitosan is the most usually applied due to its chemical structure.

Chitin (Figure 9.22), the second most abundant natural polymer in the world, is extracted from various plant and animals. However, derivations of chitin have been noticed, because of its insolubility in aqueous media. Chitosan (Figure 9.23) is deacetylated by the derivation of chitin with free amine. Unlike chitin, chitosan is soluble in diluted and organic acids. The polysaccharides are containing 2-acetamido-2-deoxy-β-D-glucose and 2-amino-2-deoxy-β-D-glucose. Deacetylation of chitin converts the acetamide groups into amino groups. The deacetylation degree (DD), one of the important effective parameters in chitosan properties has been defined as "the mole fraction of deacetylated units in the polymer chain".

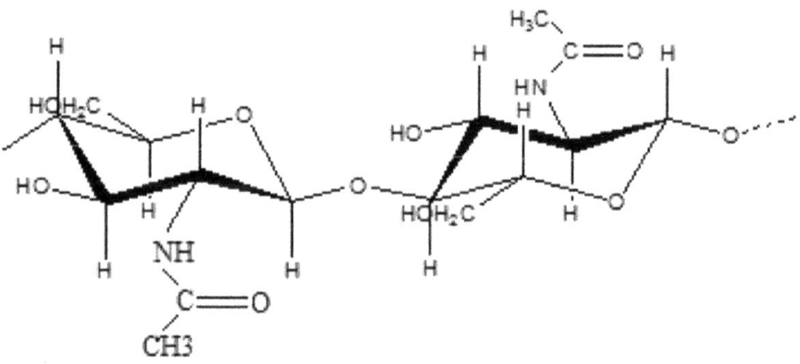

FIGURE 9.22   Structure of chitin.

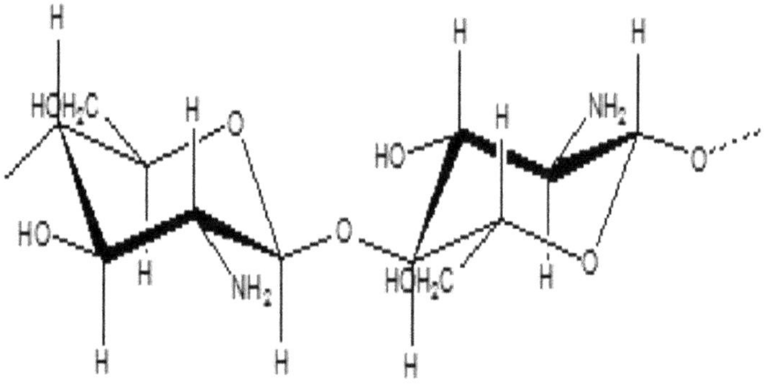

FIGURE 9.23   Structure of chitosan.

TABLE 9.6   Chiotosan applications in various fields and their main characteristics

Chitosan application		Main characteristics	Ref
Water engineering		Metal ionic adsorption	[15]
	Biosensors and immobilization of enzymes and cells	Biocompatibility, biodegradability to harmless products, non toxicity, antibacterial properties, gel forming properties and hydrophilicity, remarkable affinity to proteins	[16]
	Antimicrobial and wound dressing	Wound healing properties	[17]
	Tissue engineering	Biocompatibility, biodegradable, and antimicrobial properties	[18]
	Drug and gene delivery	Biodegradable, non toxicity, biocompatibility, high charge density, mucoadhesion	[19]
	Orthopedic or periodontal application	Antibacterial properties	[20]
Photography		Resistance to abrasion, optical characteristics, and film forming ability	[21]
Cosmetic application		Fungicidal and fungi static properties	[22]
Food preservative		Biodegradability, biocompatibility, antimicrobial activity, and non toxicity	[23]
Agriculture		Biodegradability, non toxicity, antibacterial properties, cells activator, disease, and insect resistant ability	[24]
Textile industry		Microorganism resistance and absorption of anionic dyes	[25]
Paper finishing		High density of positive charge, non toxicity, biodegradability, biocompatibility, antimicrobial, and antifungal properties	[26]
Solid-state batteries		Ionic conductivity	[27]
Chromatographic separations		Presence of free $-NH2$, primary $-OH$, and secondary $-OH$	[28]
Chitosan gel for LED and NLO applications		Dye-containing chitosan gels	[29]

The left label "Biomedical application" spans the rows from "Biosensors and immobilization of enzymes and cells" through "Orthopedic or periodontal application".

9.11.2   NANOBIOCOMPOSITES WITH CHITOSAN MATRIX

Chitosan biopolymers have a great potential in biomedical applications due to their biocompatibility and biodegradability. However, the low physical properties of chitosan limited their applications. The development of high performance chitosan biopolymers involves the incorporation of fillers that display significant mechanical reinforcement.

Polymer nanocomposites have been reinforced by nano-sized particles with high surface area to volume ratio including nanoparticles, nanoplatelets, nanofibers, and carbon nanotubes. Nowadays, carbon nanotubes are considered as potential fillers, as they improve the properties of biopolymers [31].Based on such reports, the researchers assessed the effect of CNTs fillers in the chitosan matrix and evidenced the appropriate properties of CNTs or chitosan nanobiocomposites with high potential for biomedical applications.

CARBON NANOTUBES

Carbon nanotubes, which are tubulars of Buckminster fullerene, were discovered by Iijima in 1991 [32]. They are straight segments of tube with arrangements of carbon hexagonal units. In recent years, scientists have paid great attention to CNTs, because of their superior electrical, mechanical, and thermal properties [35]. Carbon nanotubes are classified as single-walled carbon nanotubes (SWNTs) formed by a single graphene sheet, and multi-walled carbon nanotubes (MWNTs) formed by several graphene sheets wrapped around the tube core. The typical range of diameters of the carbon nanotubes are of a few nanometers (~0.8–2nm at SWNTs and ~10–400nm at MWNTs, respectively), and their lengths are up to several micrometers. There are three significant methods for synthesizing CNTs, including arc-discharge, laser ablation, and chemical vapor deposition (CVD). The production of CNTs can also be realized by other synthesis techniques such as the substrate, the sol-gel and the gas phase metal catalyst [32–48].

The C–C covalent bonding between the carbon atoms is similar to that of the graphite sheets formed by sp^2 hybridization. As a result of this structure, CNTs exhibit a high specific surface area (about 10^3) and consequently, high tensile strength (more than 200GPa) and elastic modulus (typically 1–5TPa). The carbon nanotubes also show very high thermal and electrical conductivity. However, these properties differ as to the synthesis methods employed, defects, chirality, degree of graphitization, and diameter. For instance, depending on chirality, the CNT can be metallic or semiconducting [48–52].

The preparation of CNTs solutions is impossible, because of their poor solubility. Also, a strong van der Waals interaction of CNTs among several nanotubes leads to aggregation into bundle and ropes. Therefore, various chemical and physical modification strategies will be necessary for improving their chemical affinity. There are two approaches to the surface modification of CNTs, including covalent (grafting) and non covalent bonding (wrapping) of the polymer molecule onto the surface of CNTs. In addition, the reported cytotoxic effects of CNTs *in vitro* may be mitigated by chemical surface modification. On the other hand, other studies show that the end-caps on nanotubes are more reactive than the sidewalls. Hence, the adsorption of polymers onto the CNTs surface can be utilized, together with the functionalization of defects and associated carbons [52–55].

Chemical modification of CNTs by covalent bonding is one of the important methods for improving their surface characteristics. Because of the extended π-network of the sp^2-hybridized nanotubes, CNTs have a tendency for covalent attachment, which introduces the sp^3-hybridized C atoms. These functional groups can be attached to the termini of tubes by surface-bound carboxylic acids (grafting), or by direct side-wall modifications of CNTs based on "*in situ* polymerization processing" (from grafting). Chemical functionalization of CNTs creates various activated groups (such as carboxyl, amine, fluorine, etc.) onto the CNTs surface by covalent bonds. However, there are two disadvantages for these methods. First, the CNT structure may be decomposed, due to the functionalization reaction and to a long ultrasonication process. Disruption of the π electron system is reduced as the result of these damages, leading to reduction of the electrical and mechanical properties of CNTs. Secondly, the acidic and oxidation treatments often used for the functionalization of CNTs are environmentally unfriendly. Thus, non covalent functionalization of CNTs is largely recommended, as it preserves their intrinsic properties while improving solubility and processibility. According to this method, the non covalent interaction between the π electrons of the sp^2 hybridized structure at the side-walls of CNTs and other π electrons is formed by π–π stacking. These non covalent interactions can appear between CNTs and the amphiphilic molecules (surfactants) (Figure 9.24a), polymers, and biopolymers, such as DNA, polysaccharides, and so on. In the first method, surfactants—including non ionic surfactants, anionic surfactants, and cationic surfactants—are applied for CNTs functionalization. The hydrophobic parts of the surfactants are adsorbed onto the nanotubes surface and the hydrophilic parts interact with water. Polymers and biopolymers can functionalize CNTs by two methods, including endohedral (Figure 9.24b) and wrapping (Figure 9.24c). The endohedral method is a strategy for CNTs functionalization. According to this method, nanoparticles—such as proteins and DNA—are entrapped in the inner hollow cylinders of CNTs. In another technique, the van der Waals interactions and π–π stacking between CNTs and polymer lead to wrapping of the polymer around the CNTs. Various polymers and biopolymers, such as polyaniline, DNA, and chitosan, interact physically through wrapping of the nanotube surface and π–π stacking by the solubilized polymeric chain. However, researchers created a technique for the non covalent functionalization of SWNTs most similar to π–π stacking by PPE, without polymer wrapping [53–75].

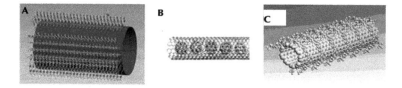

FIGURE 9.24 Non covalent functionalization of CNTs by (a) surfactants, (b) wrapping, and (c) endohedral.

Such functionalization methods can provide many applications of CNTs, one of the most important one being in biomedical science, as biosensors, drug delivery, and tissue engineering [75–78].

9.11.3 NANOTUBE COMPOSITES

Considering the low physical characteristics of biopolymers, fillers are recommended for the reinforcement of their electrical, mechanical, and thermal properties. Following the discovery of CNTs, many efforts have been made for their application as fillers in other polymers for improving the properties of the matrix polymer. First time, Ajayan was the first to apply in 1994, CNTs as a filler in epoxy resin, by the alignment method. Later on, numerous studies have focused on CNTs as excellent substitutes for conventional nanofillers in the nanocomposites. Recently, many polymers and biopolymers have been reinforced by CNTs. As already mentioned, these nanocomposites have remarkable characteristics, compared to the bulk materials, due to their unique properties [79–81].

Several parameters affect the mechanical properties of the composites, including proper dispersion, a large aspect ratio of the filler, interfacial stress transfer, a good alignment of reinforcement, and solvent selection.

The uniformity and stability of nanotube dispersion in polymer matrices are most important parameters for the performance of composites. A good dispersion leads to efficient load transfer concentration centers in composites and to uniform stress distribution. Scientists reviewed the dispersion and functionalization techniques of carbon nanotubes for polymer-based nanocomposites, as well as their effects on the properties of CNT or polymer nanocomposites. They demonstrated that the control of this two factors lead to uniform dispersion. Overall, the results showed that a proper dispersion enhanced a variety of mechanical properties of nanocomposites [71–83].

The fiber aspect ratio, defined as "the ratio of average fiber length to fiber diameter," is one of the main effective parameters on the longitudinal modulus. Generally, the carbon nanotubes have a high aspect ratio, but their ultimate performance in a polymer composite is different. The high aspect ratio of the dispersed CNTs could lead to a significant load transfer. However, aggregation of the nanotubes could lead to a decrease in the effective aspect ratio of the CNTs. Hence, the properties of nanotube composites lower are more enhanced than predicted, which is one of the processing challenges and of a poor CNTs dispersion [86].

The interfacial stress transfer has been performed by applying external stresses to the composites. The assessments showed that fillers take a significant larger share of the load, due to the CNTs-polymer matrix interaction. Also, the literature on the mechanical properties of polymer nanotube composites explains the enhancement of Young's modulus by CNTs addition. Researchers investigated the effect of stress-induced fragmentation of multi-walled carbon nanotubes in a polymer matrix. The results showed that polymer deformation generates tensile stress, which is transmitted to CNTs [84–88]

A homogeneous CNT or polymer matrix alignment in the composite is another effective parameter of carbon nanotube composites. Some researchers assessed the effects of CNT alignment on electrical conductivity and mechanical properties of SWNT or epoxy nanocomposites. The electrical conductivity, Young's modulus and tensile strength of the SWNT or epoxy composites increase with increasing SWNT alignment, due to an increased interface bonding of CNTs in the polymer matrix.

Researchers examined the effect of solvent selection on the mechanical properties of CNTs–polymer composites fabricated from double-walled nanotubes and polyvinyl alcohol composites in different solvents including water, DMSO, and NMP. This work shows that solvent selection can have a dramatic effect on the mechanical properties of CNTs-polymer composites. Also, a critical CNTs concentration was defined as an optimum improvement of the mechanical properties of nanotube composites [89–92].

Other parameters influencing the mechanical properties of nanotube composites are the size, crystallinity, crystalline orientation, purity, entanglement, and straightness. Generally, the ideal CNT properties depend on both matrix and type of application.

The various functional groups on the CNTs surface permit coupling with the polymer matrix. A strong interface between the coupled CNT or polymer creates an efficient stress transfer. As a previous point, stress transfer is a critical parameter for controlling the mechanical properties of a composite. However, the covalent treatment of CNT reduces the electrical and thermal properties of CNTs. Finally, these reductions affect the properties of nanotubes.

The matrix polymer can be wrapped around the CNT surface by non covalent functionalization, which improves the composite properties through various specific interactions. In their turn, these interactions can improve the properties of nanotube composites. In this context, researchers evaluated the electrical and thermal conductivity in CNTs or epoxy composites. Figures 9.25 and 9.26 show the electrical and, respectively, thermal conductivity in various filler contents, including carbon black (CB), double-walled carbon nanotube (DWNT) and multi-functionalization. The experimental results showed that electrical and thermal conductivity in nanocomposites are improved by the non covalent functionalization of CNTs [92–100].

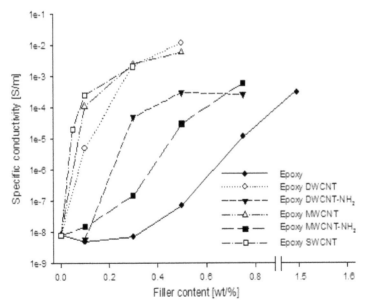

FIGURE 9.25 Electrical conductivity of nanocomposites as function of filler content in weight percent.

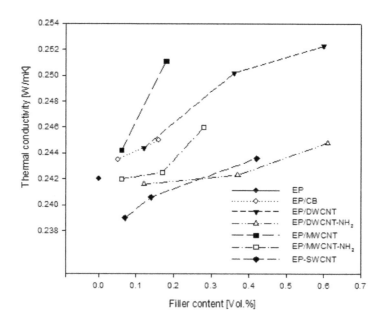

FIGURE 9.26   Thermal conductivity as function of the relative provided interfacial area per gram composite (m²/g).

9.11.4   MECHANICAL AND ELECTRICAL PROPERTIES OF NANOCOMPOSITES

Table 9.7 provides mechanical and electrical information on a CNTs or natural polymer, compared with a natural polymer. The investigations show the higher mechanical and electrical properties of CNTs or natural polymers, compared to those of natural polymers.

TABLE 9.7   Mechanical and electrical information of neat biopolymers compared to their carbon nanotube nanocomposites

Method	Biopolymer	Mechanical					Conductivity	Ref
		Tensile modulus (MPa)	Tensile strength (MPa)	Strain to failure (%)	Comparison modulus (Pa)	Stage modulus (GPa)		
Polymerized hydrogel	Neat collagen				1284 ± 94		11.37ms ± 0.16	[98]
	Collagen/CNTs				1127 ± 73		11.85ms ± 0.67	
Solution-evaporation	Neat chitosan	1.08 ± 0.04	37.7 ± 4.5				0.021ns/cm	[99, 100]
	Chitosan/CNTs	2.15 ± 0.09	74.3 ± 4.6				120ns/cm	
Wet spinning	Neat chitosan	4250						[101]
	Chitosan/CNTs	10250						
Electrospinning	Neat silk	140 ± 2.21	6.18 ± 0.3	5.78 ± 0.65			0.028s/cm	[102]
	Silk/CNTs	4817.24 ± 69.23	44.46 ± 2.1	1.22 ± 0.14			0.144s/cm	
Dry-jet wet spinning	Neat cellulose	13100 ± 1100	198 ± 25	2.8 ± 0.7		5.1	negligible	[103, 104]
	Cellulose/CNTs	14900 ± 1300	257 ± 9	5.8 ± 1.0		7.4	3000s/cm	
Electrospinning	Neat cellulose	553 ± 39	21.9 ± 1.8	8.04 ± 0.27				[105]
	Cellulose/CNT	1144 ± 37	40.7 ± 2.7	10.46 ± 0.33				

9.12 APPLICATIONS

In recent years, special attention has been paid to applying nanotube composites in various fields. Researchers reviewed nanotubes composites based on gas sensors, known as playing an important role in industry, environmental monitoring, biomedicine, and so forth. The unique geometry, morphology, and material properties of CNTs permitted their application in gas sensors.

There are many topical studies for biological and biomedical applications of carbon nanotube composites, based on their biocompatibility, such as biosensors, tissue engineering, and drug delivery in biomedical technology [100, 110].

On the other hand, the light weight, mechanical strength, electrical conductivity, and flexibility are significant properties of carbon nanotubes for aerospace applications.

Researchers overviewed carbon nanotube composite applications, including electrochemical actuation, strain sensors, power harvesting, and bioelectronic sensors, discussing the appropriate elastic and electrical properties for using nanoscale smart materials in the synthesis of intelligent electronic structures. In this context, a polyaniline or SWNTs composite fiber was previously developed, for evidencing its high strength, robustness, good conductivity, and pronounced electroactivity. The new battery materials were presented, which were made of the enhanced performance of artificial muscles by using these carbon nanotube composites.

Researchers addressed a sustainable environment and green technologies perspective for carbon nanotube applications. These contexts are including many engineering fields, such as waste water treatment, air pollution monitoring, biotechnologies, renewable energy technologies, and green nanocommmposites.

Researchers first discovered the photoinduced electron transfer from CNTs, after which the optical and photovoltaic properties of carbon nanotube composites have been studied. The results obtained suggested a possible creation of photovoltaic devices, due to the hole-collecting electrode of CNTs [110–118].

Food packaging is another remarkable application of carbon nanotube composites. Usually, poor mechanical and barrier properties have limited the application of biopolymers. Hence, an appropriate filler is necessary for promoting the matrix properties. The unique properties of CNTs have improved thermal stability, strength and modulus, as well as the water vapor transmission rate of the industrially applied composites.

In recent decades, scientists' interest for the creation of chitosan or CNTs composites is explained by their unique properties. They attempted to create new properties by adding CNTs to chitosan biopolymers. For example, several articles were published, devoted to their various applications, summarized in the graph plotted in Figure 9.27.

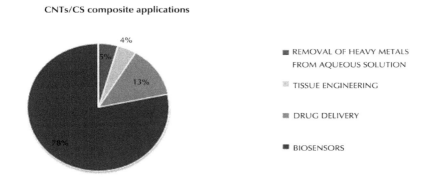

CNTs/CS composite applications

FIGURE 9.27   Chitosan or CNTs nano-composites applications.

9.13   CHITOSAN OR CARBON NANOTUBE NANOFLUIDS

The viscosity and thermal conductivity of nanofluids containing MWNTs stabilized by chitosan were investigated. The MWNTs fluid was stabilized with a chitosan solution. The investigations showed that the thermal conductivity enhancements obtained were significantly higher than those predicted by Maxwell's theory. It was also observed that dispersing chitosan into deionized water significantly increased the viscosity of nanofluid, which explains its non Newtonian behavior.

9.14   PREPARATION METHODS OF CHITOSAN OR CNTS NANO-COMPOSITES

There are several methods for the creation of nanobiocomposites, some of them considered for the preparation of chitosan or CNTs nano-composites. These methods are discussed in the following.

9.14.1   SOLUTION-CASTING-EVAPORATION

The electrochemical sensing of a carbon nanotube or chitosan system on dehydrogenase enzymes for preparing glucose biosensors was first assessed in 2004. The nanotube composite was prepared by a solution-casting-evaporation method, according to which the CNT or CHT films were prepared by casting of the CNT or CHT solution on the surface of a glassy carbon electrode, and then dried. Such a CNT or CHT system showed a new biocomposite platform for the development of dehydrogenase-based electrochemical biosensors, due to the provided signal transduction of CNT. The encouraging results of this composite utilization in biomedical applications were highly encouraging, challenging numerous other investigations.

The effect of the CNT or CS matrix on the direct electron transfer of glucose oxidase and glucose biosensor was examined by some researchers. They exhibited high sensitivity and better stability, compared with pure chitosan films. Other investigators [122] used SWNT or CS films for preparing a new galactose biosensor with highly reliable detection of galactose. Some researchers [123] immobilized lactate dehydrogenase within MWNT or CS nanocomposite for producing lactate biosensors, as they provided fast response time and high sensitivity. Also, other investigators demonstrated that immobilization of GOD molecules into a chitosan-wrapped SWNT film is an efficient method for the development of a new class of very sensitive, stable and reproducible electrochemical biosensors.

Several experiments were performed on DNA biosensor based on chitosan film doped with carbon nanotubes. The observation made was that a CNT or CHT film can be used as a stable and sensitive platform for

DNA detection. The results demonstrated improved sensor performance by adding CNT to a chitosan film. Moreover, the analytical performance of glassy carbon electrodes modified with a dispersion of MWNT or CS for DNA quantification was reported. This new platform immobilized the DNA and opened the doors to new strategies for the development of biosensors.

Other experiments reported the high sensitivity of glassy carbon electrodes modified by MWNT-CHT for cathodic stripping voltammetric measurements of bromide (Br-).

Researchers prepared an amperometric hydrogen peroxide biosensor based on composite films of MWNT/CS. The results showed the excellent electrocatalytical activity of the biosensor for H_2O_2, with good repeatability and stability.

Some researchers reported the effect of the CNT or CHT matrix on an amperometric laccase biosensor, evidencing some of its major advantages, involving detection of different substrates, high affinity and sensitivity, durable long-term stability, and facile preparation procedures [100–130].

Researchers paid particular attention to the preparation of a SWNT or CHT film by the solution-cast method, then characterized their drug delivery properties. They found out that the SWNT or CS film has enhanced the slowing down release of dexamethasone.

The growth of apatite on chitosan-multi-walled carbon nanotube composite membranes at low MWNT concentrations was also reported. Apatite was formed on composites with low concentrations.

CNT or CHT nanobiocomposites for immunosensors were produced by some researchers. In such nanobiocomposites, electron transport enhanced and improved the detection of ochratoxin-A, due to the high electrochemical properties of SWNT. Also, the CNT or CHT nanocomposite used for the detection of human chorionic gonadotrophin antibody was performed, displaying high sensitivity and good reproducibility [130–135].

9.14.2 ROPERTIES AND CHARACTERIZATION

The morphology and mechanical properties of chitosan were promoted by adding CNTs. Besides, it was demonstrated.[135] that conducting direct electron is very useful for the adsorption of hemoglobin in a CNT or CHT composite film. The studies have demonstrated that this nanobiocomposite can be used in many fields, such as biosensing and biofuel cell applications.

Some researchers evaluated the water transport behavior of chitosan porous membranes containing MWNTs. They characterized two nanotube composites with low molecular weight CSP6K and high molecular weight CSP10K. Because of the hollow nanochannel of MWNTs, located in the pore network of the chitosan membrane, the water transport results for CSP6K enhanced, when the MWNTs content exceeded a critical content. However, for CSP10K series membranes, the water transport rate decreased with increasing the MWNTs content, due to the strong compatibilizing effect of MWNTs.

Other researchers used CNT or CHT nanocomposites with poly(styrene sulfonic acid)-modified CNTs. The thermal, mechanical, and electrical properties of CNT or CHT composite films prepared by solution-casting have potential applications as separation membranes and sensor electrodes [135–140].

9.14.3 CROSSLINKING-CASTING-EVAPORATION

Researchers discovered that MWNTs can be functionalized with –COOH groups at the sidewall defects of nanotubes by carbon nanotubes.

According to a novel method, chitosan was cross-linked with free –CHO groups by glutaraldehyde and then MWNTs were added to the mixture. The cross-linked MWNT-CHT composite was immobilized with acetylcholinesterase (AChE) for detecting both acetylthiocholine and organophosphorous insecticides. On the other hand, researchers created a new method for cross-linking CHT with carboxylated CNT, involving addition of glutaraldehyde to the MWNT or CHT solution. They immobilized AChE on the composite for preparing an amperometric acetylthiocholine sensor. The suitable fabrication reproducibility, rapid response, high sensitivity, and stability obtained could provide amperometric detection of carbaryl and treazophos pesticide. Results showed the removal of heavy metals, including copper, zinc, cadmium, and nickel ions from an aqueous solution in MWNT or CHT nanocomposite films [141–145].

SURFACE DEPOSITION CROSS-LINKING

Researches decorated carbon nanotubes with chitosan by surface deposition and cross-linking processes. In this way, the chitosan macromolecules as polymer cationic surfactants were adsorbed on the CNTs surface. In this step, CHT assures a stable dispersion of CNT in an acidic aqueous solution. The pH value of the system was increased by the ammonia solution to become non dissolvable for chitosan in aqueous media. Consequently, the soluble chitosan was deposited on the surface of CNTs similar to chitosan coating. Finally, the surface-deposited chitosan was cross-linked to CNTs by glutaraldehyde for potential applications in biosensing, gene and drug delivering of this composite.

THE ELECTRO-DEPOSITION METHOD

Researchers used the nanocomposite film of CNT or CHT as a glucose biosensor by a simple and controllable method. In this one-step electrodeposition method, a pair of gold electrodes was connected to a direct current power supply and then dipped into the CNT or CHT solution. Herein, the pH near the cathode surface increased, thereby the solubility of chitosan decreased. At pH about 6.3, chitosan became insoluble and the chitosan entrapped CNT will be deposited on the cathode surface.

Other investigators characterized electrocatalytic oxidation and sensitive electroanalysis of NADH on a novel film of CS-DA-MWNTs and improved detection sensitivity. According to this new method, glutaraldehyde cross-linked CHT-DA with covalent attachment of DA molecules to the CHT chains formed by Schiff bases. Further on, the solution of MWNT dispersed in CHT-DA solution was dropped on an Au electrode for preparing CHT-DA-MWNTs film, and finally dried.

COVALENT GRAFTING

Carboxylic acid (–COOH) groups were formed on the walls of CNTs by its refluxing in an acidic solution. Carboxylated CNTs were added to the aqueous solution of chitosan. Grafting reactions were accomplished by purging CNTs or CS solution with N_2 and heating to 98°C. Researchers [146] compared the mechanical properties and water stability of CNTs-grafted-CS with the ungrafted CNTs. A significantly improved dispersion in the chitosan matrix resulted, as well as an important improvement storage modulus and water stability of the chitosan nanocomposites.

Researchers created another process for obtaining a CHT-grafted MWNT composite. By this different method, after preparing oxidized MWNT (MWNT–COOH), they generated the acyl chloride functionalized MWNT (MWNT–COCl) in a solution of thionyl chloride. In the end, the MWNT-grafted-CS was synthesized by adding CHT to the MWNT–COCl suspension in anhydrous dimethyl formamide. Covalent modification has improved interfacial bonding, resulting in a high stability of the CNT dispersion. Biosensors and other biological applications are evaluated as potential uses of this component. Others prepared a similar composite by reacting CNT–COCl and chitosan with potassium persulfate, lactic acid and acetic acid solution at 75°C. They estimated that the CNT-grafted-CHT composite can be used in bone tissue engineering, as it may improve the thermal properties [145–150].*Nucleophilic Substitution Reaction*

Covalent modification of MWNT was accomplished with low molecular weight chitosan (LMCS) [149]. This novel derivation of MWNTs can be solved in DMF, DMAc, and DMSO, but also in aqueous acetic acid solutions.

ELECTROSTATIC INTERACTION

Some authors synthesized CHT nanoparticles-coated fMWNTs composites by electrostatic interactions between the CHT particles and functionalized CNT. They prepared CHT nanoparticles and CHT microspheres by the precipitation and, respectively, crosslinking method. The electrostatic interactions between CHT particles dissolved in distilled deionized water and the carboxylated CNTs were confirmed by changing the pH of solution. The results obtained showed the same surface charges at pH 2 (both were positively charged) and pH 8 (both were negatively charged). Electrostatic interactions may occur at pH 5.5, due to different charges between the CHT particles and fCNT with positive and negative surface charges, respectively. The CHT particles or CNT composite materials could be utilized for potential biomedical applications.

Researchers prepared SWNTs or phosphotungstic acid modified SWNTs or CS composites using phosphotungstic acid as an anchor reagent to modify SWNTs. They succeed to use PW_{12}-modified SWNT with a negative surface charge and also positively charged chitosan by electrostatic interactions. These strong interfacial interactions between SWNTs and the chitosan matrix showed favorable cytocompatibility for their potential use as scaffolds for bone tissue engineering.

MICROWAVE IRRADIATION

Researchers created a new technique for the synthesis of chitosan-modified carbon nanotubes by microwave irradiation. According to this technique, solutions of MWNTs in nitric acid were placed under microwave irradiation and dried for purification of MWNTs. A mixture of purified MWNTs and chitosan solution was reacted in the microwave oven and then centrifuged. The yielded black-colored solution was adjusted to pH 8 and centrifuged for precipitation of CNT or CHT composite. This technique is much more efficient than the conventional methods.

LAYER-BY-LAYER

Researchers characterized MWNT or CHT composite rods with layer-by-layer structure prepared by the *in situ* precipitation method. The samples prepared by coating the CHT solution on the internal surface of a cylindrical tube and then filling with a MWNT or CHT solution in acetic acid. The morphological, mechanical, and thermal properties of this composite rod were also examined.

LAYER-BY-LAYER SELF ASSEMBLY

Researchers [154] produced a homogeneous multilayer film of MWNT or CHT by the layer-by-layer self-assembly method. According to this method, the negatively charged substrates were dipped into a poly (ethyleneimine) aqueous solution, a MWNTs suspension, and a CHT solution, respectively, and finally dried. In this process, both the CHT and PEI solutions were contained NaCl for the LBL assembly. The films showed stable optical properties and were appropriate for biosensors applications.

FREEZE-DRYING

Researchers synthesized and characterized a highly conductive, porous, and biocompatible MWNT or CHT biocomposite film by the freeze-drying technique. The process was preformed by freezing a MWNT or CHT dispersion into an aluminum mold, followed by drying.

WET-SPINNING

A recent report showed that chitosan is a good dispersing agent for SWNT. The authors also proposed several methods for preparing a SWNT or CHT macroscopic structure in the form of films, hydrogels, and fibers. The CNT or CHT dispersed in acetic acid was spun into a ethanol:NaOH coagulation solution bath. Better mechanical properties of wet spun fibers resulted from improved dispersion [150–156].

ELECTROSPINNING

The (CHT or MWNTs)composite nanofibers can be fabricated by electrospinning. In our experiments, different solvents including acetic acid 1–90%, formic acid, and tri-fluoroacetic acid TFA or DCM were tested for electrospinning of CHT or CNT. No jet was seen when a high voltage (even above 25kv) was applied, with 1–30% acetic acid and formic acid as solvent for CHT or CNT.

The TFA or DCM (70:30) was the only solvent that resulted during electrospinning of CHT or CNT. The scanning electron microscopic images showed homogenous fibers with an average diameter of 455nm (306–672nm), prepared by dispersing CHT or CNT in TFA or DCM 70:30. These nanofibers have potential biomedical applications.

FIGURE 9.28   Electron micrographs of electrospun fibers at chitosan concentration 10 wt.%, 24kV, 5cm, TFA or DCM: 70/30.

9.15   CASE STUDY II

It has been found that morphology of nanofibers depends on many processing parameters. These parameters can be divided into three main groups as shown in Table 9.8. Effects of "polymer concentration", "applied voltage", and "distance from needle to collector" on structure of PAN nanofiber are the particular interest of this chapter.

TABLE 9.8 Processing parameters in electrospinning

Viscosity	Solution properties
Polymer concentrationa	
Molecular weight of polymer	
Molecular weight distribution	
Electrical conductivity	
Elasticity	
Topology (branched, linear, etc.) of the polymer	
Surface tension	
Applied voltagea	
Distance from needle to collectora	
Applied voltagea	Processing conditions
Distance from needle to collectora	
Motion of target screen	
Volume feed rate	
Needle diameter	
Temperature	
Temperature	Ambient conditions
Humidity	
Atmospheric pressure	

aParameters studied in this experiment

Our electrospinning apparatus consisted of a syringe pump, a 0–50kV Dc power supply, an ammeter, and various take up devices including metal screens, belts, and other targets in an enclosed Faraday cage. The PAN fibers (thickness = 16 dtex; length = 150mm; bright) were used to prepare solutions. This polymer was dissolved in DMF solvent at 40–50°C and different concentrations.

At concentration of 8 wt.% (using 10kV and 12kV), formation of beaded fibers observed. Using 14kV and more, nanofibers without any beads were formed (Figure 9.29).

In Figure 9.29c, the "fracture points" (indicated by arrows) are due to the jet resistance to any deformation during the whipping process.

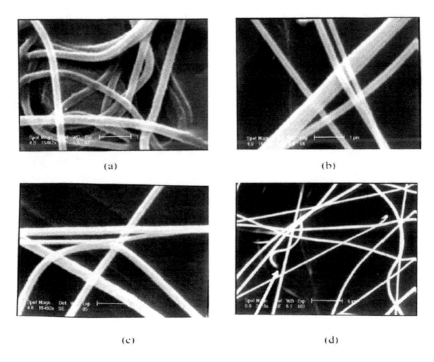

(a)   (b)

(c)   (d)

FIGURE 9.29   SEM images of nanofibers produced from 8 wt.% PAN/DMF, tip-target distance: 10cm, feed rate: 0.5ml/h, voltages used: (a) 10kv, (b) 12kv, (c) 14kv, and (d) 16kv.

At this concentration (15 wt.%), with increasing the applied voltage from 10 to 16kV and holding the tip-target distance at 10cm, the nanofibers exhibited that they could not completely dry. At 10 and 14kV, PAN webs with almost uniform diameter were obtained, but broader distributions in the diameter of nanofibers were resulted at 16kV. Fracture points were obtained at 16kV. No bead defect was present under these conditions (Figure 9.30).

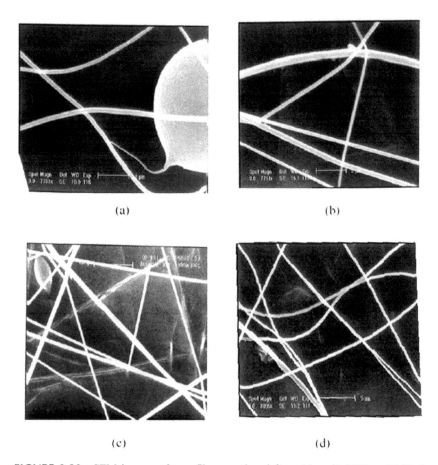

(a)   (b)

(c)   (d)

FIGURE 9.30   SEM images of nanofibers produced from 15 wt.% PAN or DMF, tip-target distance: 15cm, feed rate: 0.5ml/hr, and voltage: (a) 10kV,(b) 12kV, (c) 14kV, and (d) 16kV.

In concentration of 15 wt.%, at tip-target distance of 15cm in comparison with 10cm, the time for the solvent evaporation increased, and as a result, dry solid fibers were collected at the target. Distribution of diameter was higher at 12kV and above, fracture points and beaded fibers were formed at 14kV (Figure 9.31).

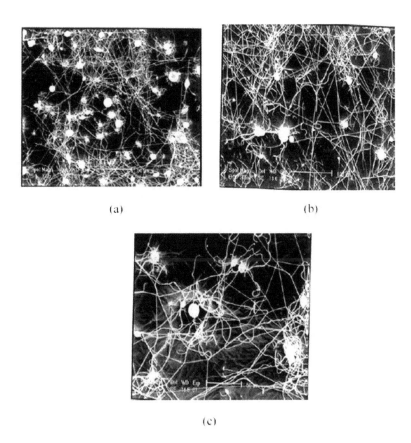

FIGURE 9.31   SEM images of beaded nanofibers produced from (a) 8 wt.%, (b) 10 wt.%, and (c) 15 wt.%, PAN/DMF. Tip-target distance: 10cm, feed rate: 0.5ml/h, and voltage: 14kv.

Drastic morphological changes were found when the concentration of polymer solution was changed. In other words, the concentration or the corresponding viscosity was one of the most effective variables to control the fiber morphology. To investigate the effect of concentration on the fibers diameter, a series of experiments were carried out in which the concentration 8, 10, 15 wt.% of PAN or DMF, with constant feeding rate of 0.5ml/hr and tip-target distance of 10cm.

The applied voltages were varied from 10 to 16kV. The effect of concentration on nanofibers diameter in these voltages are shown in Figures 9.32–9.36.

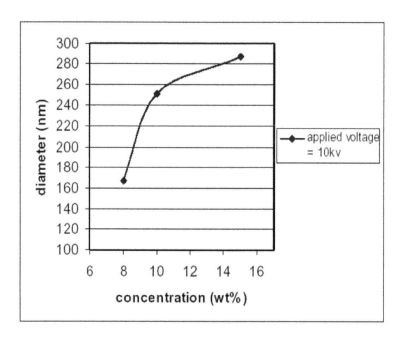

FIGURE 9.32 Effect of concentration on nanofibers diameter for applied voltage: 10kV, feed rate: 0.5ml/hr, and tip-target distance: 10cm.

The results displayed graphically in Figure 9.32 show that as the solution concentration increased, the average charge per mass ratio decreased, whereas the mean fiber diameter increased. Other considerations are the visco-elastic properties of solutions, solvent evaporation rate, details of the whipping motion, and so on.

These figures indicated the uniformly increasing trend in diameter of PAN webs with increasing the concentration at 10 and 14kV. A considerable amount of very fine nanofibers with diameters between 153 and 230nm were produced when the concentration was 8 wt.%.

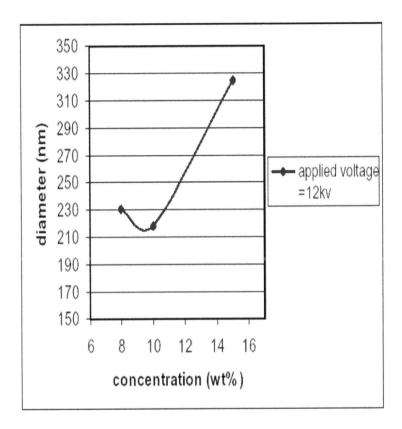

FIGURE 9.33   Effect of concentration on nanofibers diameter in applied voltage: 12kV, feed rate: 0.5ml/hr, and tip-target distance: 10cm.

In Figure 9.33, it was observed that diameter of nanofibers decreased with increasing concentration from 8 to 10 wt.% that originated from increasing the amount of stretching or elongation of the jet due to increasing the electrostatic forces at 10cm and 12kV.

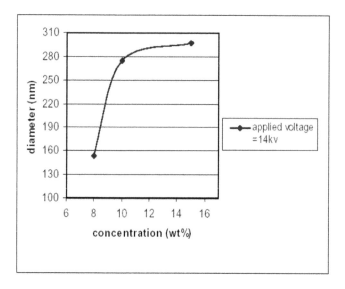

FIGURE 9.34   Effect of concentration on nanofibers diameter in applied voltage: 14kV, feed rate: 0.5ml/hr, and tip-target distance: 10cm.

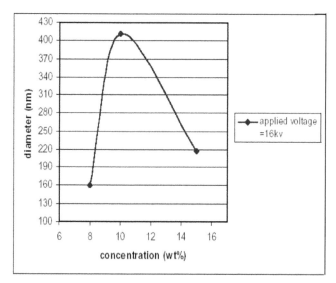

FIGURE 9.35   Effect of concentration on nanofibers diameter in applied voltage: 16kV, feed rate: 0.5ml/hr, and tip-target distance: 10cm.

As results are shown in Figures 9.35 and 9.36, there was a decrease in the average diameter of nanofibers with increasing concentration from 10 to 15 wt.%. In this case, at tip target distance of 10cm, the electric field strength is increased, so it will increase the electrostatic repulsive force on the fluid jet; on the other hand, the solution will be removed from the needle tip more quickly which favors the thinner fiber formation.

To investigate the effect of applied voltage on fiber diameter, a series of experiments carried out at voltage 10, 12, 14, and 16kV with concentration 8 to 15 wt.% , flow rate 0.5 ml/hr and tip-target distance 10cm (Figures 9.37–9.38).

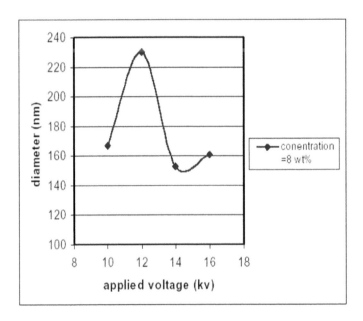

FIGURE 9.36   Effect of applied voltage on nanofibers diameter in concentration: 8 wt.%, feed rate: 0.5ml/hr, and tip-target distance: 10cm.

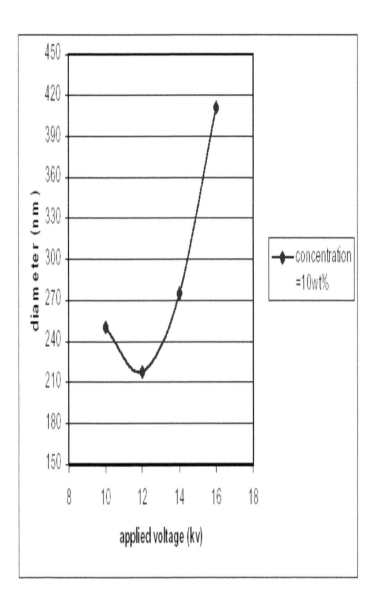

FIGURE 9.37   Effect of applied voltage on nanofibers diameter in concentration: 10 wt.%, feed rate: 0.5ml/h, tip-target distance: 10cm.

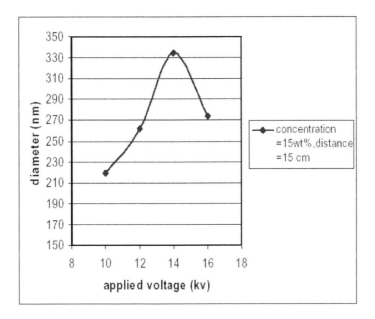

FIGURE 9.38 Effect of applied voltage on nanofibers diameter in concentration: 15 wt.%, feed rate: 0.5ml/hr, and tip-target distance: 15cm.

This increasing–decreasing or opposing trends in diameter in above figure can explain that the diameter of the jet reached a minimum after an initial increase in field strength and then became much larger with increasing electric field and with increasing the applied voltage, the electric field strength is also increased, so it will increase the electrostatic repulsive force on the jet and stretching of the jet increased, thus the thinner fiber formation took place.

9.16 CONCLUDING REMARKS

Polymer nanofibers could have many extraordinarily properties including, small diameter (and the resulting large surface area to mass ratio), highly oriented crystalline structures (and the resulting high strength), and so on. The recently fast developing technology "electrospinning" is a unique way to produce novel polymer nanofibers with diameters typically in the range of 50nm–500nm. Electrospinning of polymer nanofibers attracted significant attention during the last several years as a simple and straightforward method to produce nanostructures which are of interest in many applications. These include filter media, composite materials, biomedical application (tissue engineering, scaffolds, bandages, and drug release systems), protective clothing, optoelectronic devices, photonic crystals, and flexible photocells. The process of electrospinning is a complicated combination of polymer science, electronics, and fluid mechanics. To date, a fundamental mechanism of the process of electrospinning is still characterized only qualitatively. In this chapter, the effects of main electrospinning parameters (solution concentration, applied voltage, tip to collector distance, and volume flow rate) on AFD and CA of PAN nanofiber mat were investigated using response surface methodology. The CCD was used to establish a quantitative relationship between above factors and corresponding responses.

KEYWORDS

- **Electrospinning**
- **Electrospun**
- **Molecular weight**
- **Surface tension**
- **Viscosity**

REFERENCES

1. Praznik, W. & Stevens, C. V. (2004). *Renewable bioresources: Scope and modification for non food applications* (p. 49). Hoboken: John Wiley & Sons.
2. Parry, D. A. D. & Baker, E. N. (1984). *Reports on Progress in Physics, 47,* 1133.
3. Bhattacharyya, S., Guillot, S., Dabboue, H., Tranchant, J. -F., & Salvetat, J. -P. (2008). *Biomacromolecules, 9,* 505.
4. Lavorgna, M., Piscitelli, F., Mangiacapra, P., & Buonocore, G. G. (2010). *Carbohydrate Polymers, 82,* 291.

5. Cao, X., Chen, Y., Chang, P. R., & Huneault, M. A. (2007). *Journal of Applied Polymer Science, 106,* 1431.
6. Ramakrishna, S., Mayer, J., Wintermantel, E., & Leong, K. W. (2001). *Composites Science and Technology, 61,* 1189.
7. Liang, D., Hsiao, B. S., & Chu, B. (2007). *Advanced Drug Delivery Reviews, 59,* 1392.
8. Liu, Z., Jiao, Y., Wang, Y., Zhou, C., & Zhang, Z. (2008). *Adv. Drug Delivery Rev., 60,* 1650.
9. Agboh, O. C. & Qin, Y. (1997). *Polymers for Advanced Technologies, 8,* 355.
10. Aranaz, I., Mengíbar, M., Harris, R., Paños, I., Miralles, B., Acosta, N., Galed, G., & Heras, Á. (2009). *Current Chemical Biology, 3,* 203.
11. Zhang, Y., Xue, C., Xue, Y., Gao, R., & Zhang, X. (2005). *Carbohydrate Research, 340,* 1914.
12. VandeVord, P. J., Matthew, H. W. T., DeSilva, S. P., Mayton, L. Wu, B., & Wooley, P. H. (2002). *Journal of Biomedical Materials Research, 59,* 585.
13. Ratajska, M., Strobin, G., Wiśniewska-Wrona, M., Ciechańska, D., Struszczyk, H., Boryniec, S., Biniaś, D., & Biniaś, W. (2003). *Fibres and Textiles in Eastern Europe, 11,* 75.
14. Jayakumar, R., Prabaharan, M., Nair, S. V., & Tamura, H. (2010). *Biotechnology Advances, 28,* 142.
15. Bamgbose, J. T., Adewuyi, S., Bamgbose, O., & Adetoye, A. A. (2010). *African Journal of Biotechnology, 9,* 2560.
16. Krajewska, B. (2004). *Enzyme and Microbial Technology, 35,* 126.
17. Ueno, H., Mori, T., & Fujinaga, T. (2001). *Advanced Drug Delivery Reviews, 52,* 105.
18. Kim, I. -Y., Seo, S. -J., Moon, H. -S., Yoo, M. -K., Park, I. -Y., Kim, B. -C., & Cho, C. -S. (2008). *Biotechnology Advances, 26,* 1.
19. Sinha, V. R., Singla, A. K., Wadhawan, S., Kaushik, R., Kumria, R., Bansal, K., & Dhawan, S. (2004). *International Journal of Pharmaceutics, 274,* 1.
20. Martino, A. D., Sittinger, M., & Risbud, M. V. (2005). *Biomaterials, 26,* 5983.
21. Muzzarelli, R. A. A. (1997). *Cellular and Molecular Life Science, 53,* 131.
22. Muzzarelli, R. A. A., Muzzarelli, C., Tarsi, R., Miliani, M., Gabbanelli, F., & Cartolari, M. (2001). *Biomacromolecules, 2,* 165.
23. Dutta, P. K., Tripathi, S., Mehrotra, G. K., & Dutta, J. (2009). *Food Chemistry, 114,* 1173.
24. Boonlertnirun, S., Boonraung, C., & Suvanasara, R. (2008). *Journal of Metals, Materials and Minerals, 18,* 47.
25. Huang, K. -S., Wu, W. -J., Chen, J. -B., & Lian, H. -S. (2008). *Carbohydrate Polymers, 73,* 254.
26. Lertsutthiwong, P., Chandrkrachang, S., & Stevens, W. F. (2000). *Journal of Metals, Materials and Minerals, 10,* 43.
27. Subban, R. H. Y. & Arof, A. K. (1996). *Physica Scripta, 53,* 382.
28. Ottøy, M. H., Vårum, K. M., Christensen, B. E., Anthonsen, M. W., & Smidsrød, O. (1996). *Carbohydrate Polymers, 31,* 253.
29. Dutta, P. K., Dutta, J. & Tripathi, V. S. (2004). *Journal of Scientific and Industrial Research, 63,* 20.
30. Li, Q., Zhou, J., & Zhang, L. (2009). *Journal of Polymer Science: Polymer Physics, 47,* 1069.

31. Thostenson, E. T., Li, C., & Chou, T. -W. (2005). *Composites Science and Technology, 65,* 491.
32. Iijima, S. (1991). *Nature, 354,* 56.
33. Benthune, D. S., Kiang, C. H., Vries, M. S. d.. Gorman, G., Savoy, R., Vazquez, J., & Beyers, R. (1993). *Nature, 363,* 605.
34. Iijima, S. & Ichihashi, T. (1993). *Nature, 363,* 603.
35. Trojanowicz, M. (2006). *Trends in Analytical Chemistry, 25,* 480.
36. Duclaux, L. (2002). *Carbon, 40,* 1751.
37. Wang, Y. Y., Gupta, S., Garguilo, J. M., Liu, Z. J., Qin, L. C., & Nemanich, R. J. (2005). *Diamond and Related Materials, 14,* 714.
38. Guo, J., Datta, S., & Lundstrom, M. (2004). *IEEE Transaction on Electron Devices, 51,* 172.
39. Kuo, C. -S., Bai, A., Huang, C. -M., Li, Y. -Y., Hu, C. -C., & Chen, C. -C. (2005). *Carbon, 43,* 2760.
40. Jacobsen, R. L., Tritt, T. M., Guth, J. R., Ehrlich, A. C., & Gillespie, D. J. (1995). *Carbon, 33,* 1217.
41. Journet, C., Maser, W. K., Bernier, P., Loiseau, A., Chapelle, M. L. l., Lefrant, S., Deniard, P., Leek, R., & Fischer, J. E. (1997). *Nature, 388,* 756.
42. Thess, A., Lee, R., Nikolaev, P., Dai, H., Petit, P., Robert, J., Xu, C., Lee, Y. H., Kim, S. G., Rinzler, A. G., Colbert, D. T., Scuseria, G. E., Tomanek, D., Fischer, J. E., & Smalley R. E. (1996). *Science, 273,* 483.
43. Cassell, A. M., Raymakers, J. A., Kong, J., & Dai, H. (1999). *Journal of Physical Chemistry B, 103,* 6482.
44. Fan, S., Liang, W., Dang, H., Franklin, N., Tombler, T., Chapline, M., & Dai, H. (2000). *Physica E., 8,* 179.
45. Xie, S., Li W., Pan, Z., Chang, B., & Sun, L. (2000). *Materials Science and Engineering A, 286,* 11.
46. Tang, Z. K., Zhang, L., Wang, N., Zhang, X. X., Wen, G. H., Li, G. D., Wang, J. N., Chan, C. T., & Sheng, P. (2001). *Science, 292,* 2462.
47. Peigney, A., Laurent, C., Flahaut, E., Bacsa, R. R., & Rousset, A. (2001). *Carbon, 39,* 507.
48. Pan, Z. W., Xie, S. S., Lu, L., Chang, B. H., Sun, L. F., Zhou, W. Y., Wang, G., & Zhang, D. L. (1999). *Applied Physics Letters, 74,* 3152.
49. Forro, L., Salvetat, J. P., Bonard, J. M., Basca, R., Thomson, N. H., Garaj, S., Thien-Nga, L., Gaal, R., Kulik, A., Ruzicka, B., Degiorgi, L., Bachtold, A., Schonenberger, C., Pekker, S., & Hernadi, K. (2000). Electronic and mechanical properties of carbon nanotubes. In Tomanek & Enbody (Eds.), *Science application nanotubes* (p. 297). New York: Plenum Publishers.
50. Frank, S., Poncharal, P., Wang, Z. L., & Heer, W. A. D. (1998). *Science, 280,* 1744.
51. Britzab, D. A. & Khlobystov, A. N. (2006). *Chemical Society Reviews, 35,* 637.
52. Andrews, R. & Weisenberger, M. C. (2004). *Current Opinion in Solid State and Materials Science, 8,* 31.
53. Hirsch, A. (2002). *Angewandte Chemie International Edition, 41,* 1853.
54. Firme, C. P. & Prabhakar, R. B. (2010). *Nanomedicine: Nanotechnology, Biology and Medicine, 6,* 245.

55. Niyogi, S., Hamon, M. A., Hu, H., Zhao, B., Bhowmik, P., Sen, R., Itkis, M. E., & Haddon, R. C. (2002). *Accounts of Chemical Research, 35*, 1105.
56. Kuzmany, H., Kukovecz, A., Simona, F., Holzweber, M., Kramberger, C., & Pichler, T. (2004). *Synthetic Metals, 141*, 113.
57. Spitalsky, Z., Tasis, D., Papagelis, K., & Galiotis, C. (2010). *Progress in Polymer Science, 35*, 357.
58. Narain, R., Housni, A., & Lane, L. (2006). *Journal of Polymer Science Part A: Polymer Chemistry, 44*, 6558.
59. Wang, M., Pramoda, K. P., & Goh, S. H. (2006). *Carbon, 44*, 613.
60. Touhara, H., Yonemoto, A., Yamamoto, K., Komiyama, S., Kawasaki, S., Okino, F. Yanagisawa, T., & Endo, M. (2002). *Fluorine Chemistry, 114*, 181.
61. Zhang, W., Sprafke, J. K., Ma, M., Tsui, E. Y., Sydlik, S. A., Rutledge, G., & Swager, T. M. (2009). *Journal of the American Chemical Society, 131*, 8446.
62. He, P., Gao, Y., Lian, J., Wang, L., Qian, D., Zhao, J., Wang, W., Schulz, M. J., Zhou, X. P., & Shi, D. (2006). *Composites Part A: Applied Science and Manufacturing, 37*, 1270.
63. Sulong, A. B., Azhari, C. H., Zulkifli, R., Othman, M. R., & Park, J. (2009). *European Journal of Scientific Research, 33*, 295.
64. Wang, C., Guo, Z. X., Fu, S., Wu, W., & Zhu, D. (2004). *Progress in Polymer Science, 29*, 1079.
65. Rausch, J., Zhuang, R. -C., & Mäder, E. (2010). *Composites Part A: Applied Science and Manufacturing, 41*, 1038.
66. Sahoo, N. G., Rana, S., Cho, J. W., Li, L., & Chan, S. H. (2010). *Progress in Polymer Science, 35*, 837.
67. Zheng, D., Li, X., & Ye, J. (2009). *Bioelectrochemistry, 74*, 240.
68. Zhang, X., Meng, L., & Lu, Q. (2009). *ACS Nano, 10*, 3200.
69. Wang, H. (2009). *Current Opinion in Colloid and Interface Science, 14*, 364.
70. Schlüter, A. D., Hirsch, A., & Vostrowsky, O. (2005). *Functional Molecular Nanostructures, 245*, 193.
71. Ma, P. -C., Siddiqui, N. A., Marom, G., & Kim, J. -K. (2010). *Composites Part A: Applied Science and Manufacturing, 41*, 1345.
72. Mottaghitalab, V., Spinks, G. M., & Wallace, G. G. (2005). *Synthetic Metals, 152*, 77.
73. Cheung, W., Pontoriero, F., Taratula, O., Chen, A. M., & He, H. (2010). *Advanced Drug Delivery Reviews, 62*, 633.
74. Piovesan, S., Cox, P. A., Smith, J. R., Fatouros, D. G., & Roldo, M. (2010). *Physical Chemistry Chemical Physics, 12*, 15636.
75. Chen, J., Liu, H., Weimer, W. A., Halls, M. D., Waldeck, D. H., & Walker, G. C. (2002). *Journal of the American Chemical Society, 124*, 9034.
76. Vamvakaki, V., Fouskaki, M., & Chaniotakis, N. (2007). *Analytical Letters, 40*, 2271.
77. Kang, Y., Liu, Y. -C., Wang, Q., Shen, J. -W., Wu, T., & Guan, W. -J. (2009). *Biomaterials, 30*, 2807.
78. Harrison, B. S. & Atala, A. (2007). *Biomaterials, 28*, 344.
79. Moniruzzaman, M. & Winey, K. I. (2006). *Macromolecules, 39*, 5194.
80. Ajayan, P. M., Stephan, O., Colliex, C., & Trauth, D. (1994). *Science, 265*, 1212.
82. Liu, P. (2005). *European Polymer Journal, 41*, 2693.

83. Manchado, M. A. L., Valentini, L., Biagiotti, J., & Kenny, J. M. (2005). *Carbon, 43,* 1499.
84. Wang, Q. & Varadan, V. K. (2005). *Smart Materials and Structures, 14,* 281.
85. Mallick, P. K. (2008). *Fiber reinforced composites: Materials, manufacturing, and design* (p. 43). London: Taylor & Francis Group.
86. Shaffer, M. S. P. & Sandler, J. K. W. (2006). *Processing and properties of nanocomposites* (p. 31). Singapore: World Scientific Publishing.
87. Nan, C. W., Shi, Z., & Lin, Y. (2003). *Chemical Physics Letters, 375,* 666.
88. Coleman, J. N., Khan, U., Blau, W. J., & Gun'ko, Y. K. (2006). *Carbon, 44,* 1624.
89. Wagner, H. D., Lourie, O., Feldman, Y., & Tenne, R. (1998). *Applied Physics Letters, 72,* 188.
90. Wang, Q., Dai, J., Li, W., Wei, Z., & Jiang, J. (2008). *Composites Science and Technology, 68,* 1644.
91. Khan, U., Ryan, K., Blau, W. J., & Coleman, J. N. (2007). *Composites Science and Technology, 67,* 3158.
92. Allaoui, A., Bai, S., Cheng, H. M., & Bai, J. B. (2002). *Composites Science and Technology, 62,* 1993.
93. Esawi, A. M. K. & Farag, M. M. (2007). *Materials and Design, 28,* 2394.
94. Gojny, F. H., Wichmann, M. H. G., Fiedler, B., Kinloch, I. A., Bauhofer, W., Windle, A. H., & Schulte, K. (2006). *Polymer, 47,* 2036.
95. Kamaras, K., Itkis, M. E., Hu, H., Zhao, B., & Haddon, R. C. (2003). *Science, 301,* 1501.
96. Shenogin, S., Bodapati, A., Xue, L., Ozisik, R., & Keblinski, P. (2004). *Applied Physics Letters, 85,* 2229.
97. MacDonald, R. A., Laurenzi, B. F., Viswanathan, G., Ajayan, P. M., & Stegemann, J. P. (2005). *Journal of Biomedical Materials Research Part A, 74A,* 489.
98. Bose, S., Khare, R. A., & Moldenaers, P. (2010). *Polymer, 51,* 975.
99. Tosun, Z. & McFetridge, P. S. (2010). *Journal of Neural Engineering, 7,* 1.
100. Wang, S. -F., Shen, L., Zhang, W. -D., & Tong, Y. -J. (2005). *Biomacromolecules, 6,* 3067.
101. Spinks, G. M., Shin, S. R., Wallace, G. G., Whitten, P. G., Kim, S. I., & Kim, S. J. (2006). *Sensors and Actuators B, 115,* 678.
102. Gandhi, M., Yang, H., Shor, L., & Ko, F. (2009). *Polymer, 50,* 1918.
103. Rahatekar, S. S., Rasheed, A., Jain, R., Zammarano, M., Koziol, K. K., Windle, A. H., Gilman, J. W., & Kumar, S. (2009), *Polymer, 50,* 4577.
104. Zhang, H., Wang, Z., Zhang, Z., Wu, J., Zhang, J., He, J. (2007). *Advanced Materials, 19,* 698.
105. Lu, P. & Hsieh, Y. -L. (2010). *ACS Applied Materials and Interfaces, 2,* 2413.
106. Wang, Y. & Yeow, J. T. W. (2009). *Journal of Sensors, 2009,* 1.
107. Yang, W., Thordarson, P., Gooding, J. J., Ringer, S. P., & Braet, F. (2007). *Nanotechnology, 18,* 1.
108. Wang, J. (2005). *Electroanalysis, 17,* 7.
109. Foldvari, M. & Bagonluri, M. (2008). *Nanomedicine: Nanotechnology, Biology and Medicine, 4,* 183.
110. Belluccia, S., Balasubramanianab, C., Micciullaac, F., & Rinaldid, G. (2007). *Journal of Experimental Nanoscience, 2,* 193.

111. Kang, I., Heung, Y. Y., Kim, J. H., Lee, J. W., Gollapudi, R., Subramaniam, S., Narasimhadevara, S., Hurd, D., Kirikera, G. R., Shanov, V., Schulz, M. J., Shi, D., Boerio, J., Mall, S., & Ruggles-Wren, M. (2006). *Composites Part B, 37,* 382.
112. Mottaghitalab, V., Spinks, G. M., & Wallace, G. G. (2006). *Polymer, 47,* 4996.
113. Wang, C. Y., Mottaghitalab, V., Too, C. O., Spinks, G. M., & Wallace, G. G. (2007). *Journal of Power Sources, 163,* 1105.
114. Mottaghitalab, V., Xi, B., Spinks, G. M., & Wallace, G. G. (2006). *Synthetic Metals, 156,* 796.
115. Ong, Y. T., Ahmad, A. L., Zein, S. H. S., & Tan, S. H. (2010). *Brazilian Journal of Chemical Engineering, 27,* 227.
116. Sariciftci, N. S., Smilowitz, L., Heeger, A. J., & Wudi, F. (1992). *Science, 258,* 1474.
117. Harris, P. J. F. (2004). *International Metallurgical Reviews, 49,* 31.
118. Azeredo, H. M. C. d. (2009). *Food Research International, 42,* 1240.
119. Phuoc, T. X., Massoudi, M., & Chen, R. -H. (2011). *International Journal of Thermal Sciences, 50,* 12.
120. Zhang, M., Smith, A., & Gorski, W. (2004). *Analytical Chemistry, 76,* 5045.
121. Liu, Y., Wang, M., Zhao, F., Xu, Z., & Dong, S. (2005). *Biosensors and Bioelectronics, 21,* 984.
122. Tkac, J., Whittaker, J. W., & Ruzgas, T. (2007). *Biosensors and Bioelectronics, 22,* 1820.
123. Tsai, Y. -C., Chen, S. -Y., & Liaw, H. -W. (2007). *Sensors and Actuators B, 125,* 474.
124. Zhou, Y., Yang, H., & Chen, H. -Y. (2008). *Talanta, 76,* 419.
125. Li, J., Liu, Q., Liu, Y., Liu, S., & Yao, S. (2005). *Analytical Biochemistry, 346,* 107.
126. Bollo, S., Ferreyr, N. F., & Rivasb, G. A. (2007). *Electroanalysis, 19,* 833.
127. Zeng, Y., Zhu, Z. -H., Wang, R. -X., & Lu, G. -H. (2005). *Electrochimica Acta, 51,* 649.
128. Qian, L. & Yang, X. (2006). *Talanta, 68,* 721.
129. Liu, Y., Qu, X., Guo, H., Chen, H., Liu, B., & Dong, S. (2006). *Biosensors and Bioelectronics, 21,* 2195.
130. Naficy, S., Razal, J. M., Spinks, G. M., & Wallace, G. G. (2009). *Sensors and Actuators A, 155,* 120.
131. Yang, J., Yao, Z., Tang, C., Darvell, B. W., Zhang, H., Pan, L., Liu, J., & Chen, Z. (2009). *Applied Surface Science, 255,* 8551.
132. Kaushik, A., Solanki, P. R., Pandey, M. K., Kaneto, K., Ahmad, S., & Malhotra, B. D. (2010). *Thin Solid Films, 519,* 1160.
133. Yang, H., Yuan, R., Chai, Y., & Ying, Z. (2011). *Colloids and Surfaces B, 82,* 463.
134. Zheng, W., Chen, Y. Q., & Zheng, Y. F. (2008). *Applied Surface Science, 255,* 571.
135. Tang, C., Zhang, Q., Wang, K., Fu, Q., & Zhang, C. (2009). *Journal of Membrane Science, 337,* 240.
136. Liu, Y. -L., Chen, W. -H., & Chang, Y.-H. (2009). *Carbohydrate Polymers, 76,* 232.
137. Ghica, M. E., Pauliukaite, R., Fatibello-Filho, O., & Brett, C. M. A. (2009). *Sensors and Actuators B, 142,* 308.
138. Kandimalla, V. B. & Ju, H. (2006). *Chemistry: A European Journal, 12,* 1074.
139. Du, D., Huang, X., Cai, J., Zhang, A., Ding, J., & Chen, S. (2007). *Analytical and Bioanalytical Chemistry, 387,* 1059.
140. Du, D., Huang, X., Cai, J., & Zhang, A. (2007). *Sensors and Actuators B, 127,* 531.

141. Salam, M. A., Makki, M. S. I., & Abdelaal, M. Y. A. (2010). *Journal of Alloys and Compounds, 509,* 2582.

142. Liu, Y., Tang, J., Chen, X., & Xin, J. H. (2005). *Carbon, 43,* 3178.

143. Luo, X. -L., Xu, J. -J., Wang, J. -L., & Chen, H. -Y. (2005). *Chemical Communications, 16,* 2169.

144. Ge, B., Tan, Y., Xie, Q., Ma, M., & Yao, S. (2009). *Sensors and Actuators B, 137,* 547.

145. Shieh, Y. -T. & Yang, Y. -F. (2006). *European Polymer Journal, 42,* 3162.

146. Wu, Z., Feng, W., Feng, Y., Liu, Q., Xu, X., Sekino, T., Fujii, A., & Ozaki, M. (2007). *Carbon, 45,* 1212.

147. Carson, L., Kelly-Brown, C., Stewart, M., Oki, A., Regisford, G., Luo, Z., & Bakhmutov, V. I. (2009). *Materials Letters, 63,* 617.

148. Ke, G., Guan, W. C., Tang, C. Y., Hu, Z., Guan, W. J., Zeng, D. L., & Deng, F. (2007). *Chinese Chemical Letters, 18,* 361.

149. Baek, S. -H., Kim, B., & Suh, K. -D. (2008). *Colloids and Surfaces A, 316,* 292.

150. Zhao, Q., Yin, J., Feng, X., Shi, Z., Ge, Z., & Jin, Z. (2010) *Journal of Nanoscience and Nanotechnology, 10,* 1.

151. Yu, J. -G., Huang, K. -L., Tang, J. -C., Yang, Q., & Huang, D. -S. (2009). *International Journal of Biological Macromolecules, 44,* 316.

152. Wang, Z. -K., Hu, Q. -L., & Cai, L. (2010). *Chinese Journal of Polymer Science, 28,* 801.

153. Li, X. -B. & Jiang, X. -Y. (2010). *New Carbon Mater, 25,* 237.

154. Lau, C. & Cooney, M. J. (2008). *Langmuir, 24,* 7004.

155. Jennings, J. A., Haggard, W. O., & Bumgardner, J. D. (2010). US Patent Application, 0266694.A1.

156. Razal, J. M., Gilmore, K. J., & Wallace, G. G. (2008). *Advanced Functional Materials, 18,* 61.

157. Lynam, C., Moulton, S. E., & Wallace, G. G. (2007). *Advanced Materials, 19,* 1244.

CHAPTER 10

GREEN NANOFIBERS

A. K. HAGHI, V. MOTTAGHITALAB, M. HASANZADEH, and
B. H. MOGHADAM

CONTENTS

10.1   Introduction 273
10.2   Experimental 276
10.3   Preparation of Cht-Mwnts Dispersions 276
10.4   Electrospinning Process 277
10.5   Measurements and Characterizations Method 278
10.6   Results 279
10.7   Silk Fibers 287
      10.7.1   Preparation of Regenerated SF Solution 290
      10.7.2   Preparation of the Spinning Solution 290
      10.7.3   Electrospinning 290
      10.7.4   Characterization 291
      10.7.5   Results and Discussion 291
10.8   Modeling 301
10.9   Concuding Remarks 305
Appendix-A 306
      Curious Facts 308
      Curious Facts 309
      Curious Facts 313
APPENDIX-B 350

Appendix-C 353
Appendix-D 369
Keywords 380
References 381

10.1 INTRODUCTION

Over the recent decades, the fabrication of biodegradable nanofibers for many biomedical applications, such as tissue engineering, drug delivery, wound dressing, enzyme immobilization, and so on, has been recorded. Nanofiber fabrics have unique characteristics, such as very large surface area, ease of functionalization for various purposes and superior mechanical properties. Electrospinning (Figure 10.1) is an important technique that can be used for the production of polymer nanofibers with diameters from several micrometers down to tens of nanometers. In electrospinning, the charged jets of a polymer solution, collected on a target, are created by using an electrostatic force. Many parameters can influence the quality of fibers, including the solution properties (polymer concentration, solvent volatility and solution conductivity), the governing variables (flow rate, voltage and tip-to-collector distance), and the ambient parameters (humidity, solution temperature, air velocity in the electrospinning chamber) [1-12].

In recent years, scientists have manifested increased interest in electrospinning of natural materials, such as collagen, fibrogen, gelatin, silk, chitin and chitosan, due to their high biocompatible and biodegradable properties. Chitin is the second most abundant natural polymer in the world and chitosan (poly-(1-4)-2-amino-2-deoxy-β-D-glucose) is the deacetylated product of chitin.

Researchers are interested in this natural polymer because of its properties, including solid-state structure and chain conformations in dissolved state [13-20].

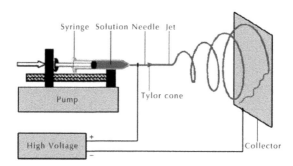

FIGURE 10.1   A typical image of electrospinning process.

The physical characteristics of electrospun nanofibers such as fiber diameter depend on various parameters which are mainly divided into three categories: solution properties (solution viscosity, solution concentration, polymer molecular weight, and surface tension), processing conditions (applied voltage, volume flow rate, spinning distance, and needle diameter), and ambient conditions (temperature, humidity, and atmosphere pressure). Numerous applications require nanofibers with desired properties suggesting the importance of the process control. It does not come true unless having a comprehensive outlook of the process and quantitative study of the effects of governing parameters [14-26].

Besides physical characteristics, medical scientists showed a remarkable attention to biocompatibility and biodegradability of nanofibers made of biopolymers such as collagen, fibrogen, gelatin, silk, chitin and chitosan. Chitin is the second abundant natural polymer in the world and Chitosan (poly-(1-4)-2-amino-2-deoxy-β-D-glucose) is the deacetylated product of chitin. CHT is well known for its biocompatible and biodegradable properties.

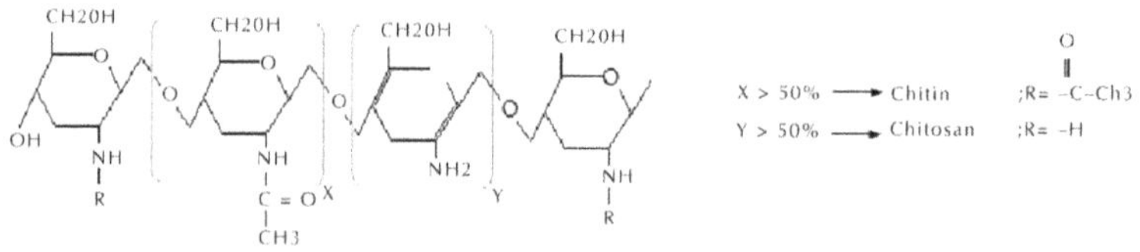

Scheme1   Chemical structures of Chitin and Chitosan bi-opolymers

Chitosan (CHT) is insoluble in water, alkali, and most mineral acidic systems. However, though its solubility in inorganic acids is quite limited, CHT is in fact soluble in organic acids, such as dilute aqueous acetic, formic, and lactic acids. CHT also has free amino groups, which make it a positively charged polyelectrolyte. This property makes CHT solutions highly viscous and complicates its electrospinning. Furthermore, the formation of strong hydrogen bonds in a 3-D network prevents the movement of polymeric chains exposed to the electrical field [27-31].

Different strategies were used for bringing CHT in nanofiber form. The three top most abundant techniques includes blending of favorite polymers for electrospinning process with CHT matrix, alkali treatment of CHT backbone to improve electro spinnability through reducing viscosity and employment of concentrated organic acid solution to produce nanofibers by decreasing of surface tension. Electrospinning of Polyethylene oxide (PEO)/CHT and polyvinyl alcohol (PVA)/CHT blended nanofiber are two recent studies based on first strategy. In the second protocol, the molecular weight of CHT decreases through alkali treatment. Solutions of the treated CHT in aqueous 70–90% acetic acid have been employed to produce nanofibers with appropriate quality and processing stability.

Using concentrated organic acids such as acetic acid and triflouroacetic acid (TFA) with and without dichloromethane (DCM) have been reported exclusively for producing neat CHT nanofibers. They similarly reported the decreasing of surface tension and at the same time enhancement of charge density of CHT solution without significant effect on viscosity. This new method suggests significant influence of the concentrated acid solution on the reducing of the applied field required for electrospinning.

The mechanical and electrical properties of neat CHT electrospun natural nanofiber mat can be improved by addition of the synthetic materials including carbon nanotubes (CNTs). The CNTs are one of the key synthetic polymers that were discovered in 1991. The CNTs either single walled nanotubes (SWNTs) or multiwalled nanotubes (MWNTs) combine the physical properties of diamond and graphite. They are extremely thermally conductive like diamond and appreciably electrically conductive like graphite. Moreover, the flexibility and exceptional specific surface area to mass ratio can be considered as significant properties of CNTs. The scientists are becoming more interested to CNTs for existence of exclusive properties such as superb conductivity and mechanical strength for various applications. To best of our knowledge, there has been no report on electrospinning of CHT/MWNTs blend, except those ones that use PVA to improve spinnability. Results showed uniform and porous morphology of the electrospun nanofibers. Despite adequate spinnability, total removing of PVA from nanofiber structure to form conductive substrate is not feasible. Moreover, thermal or alkali solution treatment of CHT/PVA/MWNTs nanofibers extremely influence on the structural morphology and mechanical stiffness. The CHT/CNT composite can be produced by the hydrogen bonds due to hydrophilic positively charged polycation of CHT due to amino groups and hydrophobic negatively charged of CNT due to carboxyl and hydroxyl groups.

10.2   EXPERIMENTAL

The CHT with degree of deacetylation of 85% and molecular weight of 5 $\times 10^5$ was supplied by Sigma-Aldrich. The MWNTs, supplied by Nutrino, have an average diameter of 4nm and the purity of about 98%. All of the other solvents and chemicals were commercially available and used as received without further purification.

10.3   PREPARATION OF CHT-MWNTS DISPERSIONS

A Branson Sonifier 250 operated at 30W used to prepare the MWNTs dispersions in CHT /organic acid (90%w/w acetic acid, 70/30 TFA/DCM) solution based on different protocols. In first approach, 3mg of as received MWNTs was dispersed into deionized water or DCM using solution sonicating for 10min (current work, sample 1). Different amount of CHT was then added to MWNTs dispersion for preparation of a 8–12 w/w% solution and then sonicated for another 5min. Figure 10.2 shows two different protocols used in this study.

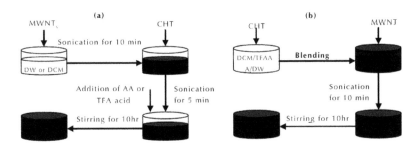

FIGURE 10.2 Two protocols used in this study for preparation of MWNTs/CHT dispersion (a) Current study (b) (*AA/DW abbreviated for acetic acid diluted in de mineralized water*).

In next step, the organic acid solution added to obtain a CHT/MWNT solution with total volume of 5mL and finally the dispersion was stirred for another 10hrs. The sample 2 was prepared using the second technique. Same amount of MWNTs were dispersed in CHT solution, and the blend with total volume of 5mL were sonicated for 10min and dispersion was stirred for 10hr.

10.4 ELECTROSPINNING PROCESS

After the preparation of spinning solution, it transferred to a 5ml syringe and became ready for spinning of nanofibers. The experiments were carried out on a horizontal electrospinning setup shown schematically in Figure 10.1. The syringe containing CHT/MWNTs solution was placed on a syringe pump (New Era NE-100) used to dispense the solution at a controlled rate. A high voltage DC power supply (Gamma High Voltage ES-30) employed to generate the required electric field for electrospinning. The positive electrode and the grounding electrode of the high voltage supplier attaches respectively to the syringe needle and flat collector wrapped with aluminum foil where electrospun nanofibers accumulates via an alligator clip to form a nonwoven mat. The voltage and the tip-to-collector distance fixed respectively on 18–24kV and 4–10cm. In addition, the electrospinning carried out at room temperature and the aluminum foil removed from the collector.

10.5   MEASUREMENTS AND CHARACTERIZATIONS METHOD

A small piece of mat placed on the sample holder and gold sputter-coated (Bal-Tec). Thereafter, the micrograph of electrospun CHT/MWNTs nanofibers was obtained using scanning electron microscope (SEM, Phillips XL-30). Fourier transform infrared spectra (FT-IR) recorded using a Nicolet 560 spectrometer to investigate the interaction between CHT and MWNT in the range of 800–4,000cm^{-1} under a transmission mode. The size distribution of the dispersed solution was evaluated using dynamic light scattering technique (Zetasizer, Malvern Instruments). The conductivity of nanofiber samples was measured using a homemade four-probe electrical conductivity cell operating at constant humidity. The electrodes were circular pins with separation distance of 0.33cm and fibers connect to pins by silver paint (SPI). Between the two outer electrodes, a constant DC current applies by Potentiostat/Galvanostat model 363 (Princeton Applied Research). The generated potential difference between the inner electrodes and the current flow between the outer electrodes recorded by digital multimeter 34401A (Agilent). Figure 10.3 illustrates the experimental setup for conductivity measurement. The conductivity (δ: S/cm) of the nanofiber with rectangular surface can then be calculated according to Equation 10.1 which parameters call for length (L:cm), width(W:cm), thickness(t:cm), DC current applied (mA) and the potential drop across the two inner electrodes(mV). All measuring repeated at least 5 times for each set of samples.

$$\delta = \frac{I \times L}{V \times W \times t} \qquad (10.1)$$

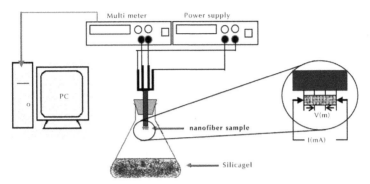

FIGURE 10.3   The experimental setup for four-probe electrical conductivity measurement of nanofiber thin film.

10.6   RESULTS

Utilisation of MWNTs in biopolymer matrix initially requires their homogenous dispersion in a solvent or polymer matrix. Dynamic light scattering (DLS) is a sophisticated technique used for evaluation of particle size distribution. DLS provides many advantages for particle size analysis to measures a large population of particles in a very short time period, with no manipulation of the surrounding medium. DLS of MWNTs dispersions indicate that the hydrodynamic diameter of the nanotube bundles is between 150 and 400nm after 10min of sonication for sample 2 (Figure 10.4).

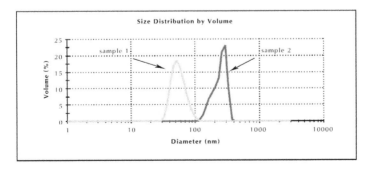

FIGURE 10.4   Hydrodynamic diameter distribution of MWNT bundles in CHT/acetic acid (1%) solution for different preparation technique.

The MWNTs bundle in sample 1(different approach but same sonica-tion time compared to sample 2) shows a range of hydrodynamic diameter between 20 and 100nm (Figure 10.4). The lower range of hydrodynamic diameter for sample 1 correlates to more exfoliated and highly stable nano-tubes strands in CHT solution. The higher stability of sample 1 compared to sample 2 over a long period of time was confirmed by solution stability test. The results presented in Figure 10.5 indicate that procedure employed for preparation of sample 1(current work) was an effective method for dispersing MWNTs in CHT or acetic acid solution. However, MWNTs bundles in sample 2 showed re-agglomeration upon standing after sonica-tion.

Despite the method reported in ref 35 neither sedimentation nor ag-gregation of the MWNTs bundles observed in first sample. Presumably, this behavior in sample 1 can be attributed to contribution of CHT biopol-ymer to forms an effective barrier against re-agglomeration of MWNTs nanoparticles. In fact, using sonication energy, in first step without pres-ence of solvent, make very tiny exfoliated but unstable particle in water as dispersant. Instantaneous addition of acetic acid as solvent and long mix-ing most likely helps the wrapping of MWNTs strands with CHT polymer chain.

Figure 10.6 shows the FT-IR spectra of neat CHT solution and CHT/MWNTs dispersions prepared using strategies explained in experimen-tal part. The interaction between the functional group associated with MWNTs and CHT in dispersed form has been understood through rec-ognition of functional groups. The enhanced peaks at ~1,600cm^{-1} can be attributed to (N – H) band and (C = O) band of amide functional group. However, the intensity of amide group for CHT/MWNTs dispersion increases presumably due to contribution of G band in MWNTs. More interestingly, in this region, the FT-IR spectra of MWNTs-CHT disper-sion (sample 1) have been highly intensified compared to sample 2. It correlates to higher chemical interaction between acid functionalized C – C group of MWNTs and amide functional group in CHT.

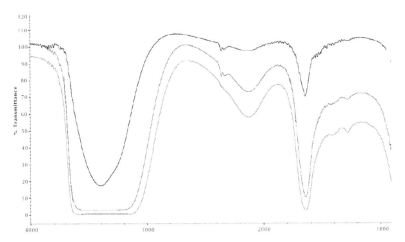

FIGURE 10.5   FT-IR spectra of CHT-MWCNT in 1% acetic acid with different techniques of dispersion

This probably is the main reason of the higher stability and lower MWNTs dimension demonstrated in Figure 10.4 and 10.5. Moreover, the intensity of protonated secondary amine absorbance at 2,400cm^{-1} for sample 1 prepared by new technique is negligible compared to sample 2 and neat CHT. Furthermore, the peak at 2,123cm^{-1} is a characteristic band of the primary amine salt, which is associated to the interaction between positively charged hydrogen of acetic acid and amino residues of CHT. In addition, the broad peaks at ~3,410cm^{-1} due to the stretching vibration of OH group superimposed on NH stretching bond and broaden due to hydrogen bonds of polysaccharides. The broadest peak of hydrogen bonds observed at 3,137–3,588cm^{-1} for MWNTs/CHT dispersion prepared by new technique (sample1).

The different solvents including acetic acid 1–90%, pure formic acid, and TFA/DCM tested for preparation of spinning solution-using protocol explained for sample 1. Upon applying of the high voltage even above 25kV no polymer jet forms using of acetic acid 1–30% and formic acid as the solvent for CHT/MWNT. However, experimental observation shows bead formation when the acetic acid (30–90%) used as the solvent. Therefore, one does not expect the formation of electrospun fiber of CHT/MWNTs using prescribed solvents (data not shown).

Figure 10.6 shows scanning electronic micrographs of the CHT/ MWNTs electrospun nanofibers in different concentration of CHT in TFA/ DCM (70:30) solvent. As presented in Figure 10.6a, at low concentrations of CHT the beads deposited on the collector and thin fibers co-exited among the beads. When the concentration of CHT increases as shown in Figure 10.6a–c the bead density decreases. Figure 10.6c show the homogenous electrospun nanofibers with minimum beads, thin and interconnected fibers. More increasing of concentration of CHT lead to increasing of interconnected fibers at Figure 10.6d–e. Figure 10.7 shows the effect of concentration on average diameter of MWNTs/CHT electrospun nanofibers. Our assessments indicate that the fiber diameter of CHT/MWNTs increases with the increasing of the CHT concentration. Hence, CHT/ MWNTs (10 wt.%) solution in TFA/DCM (70:30) considered as resulted in optimized concentration. An average diameter of 275nm (Figure 10.6c: diameter distribution, 148–385) investigated for this conditions.

| | Magnification | | Fiber diameter (nm) |
	5,000x	10,000x	
8%	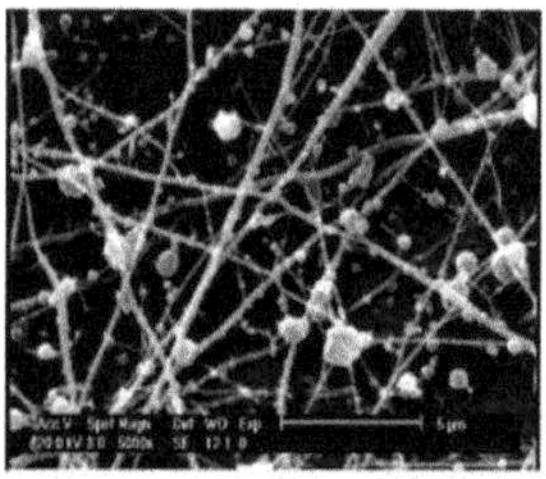		Max: 277 Min: 70 Avg: 137
9%	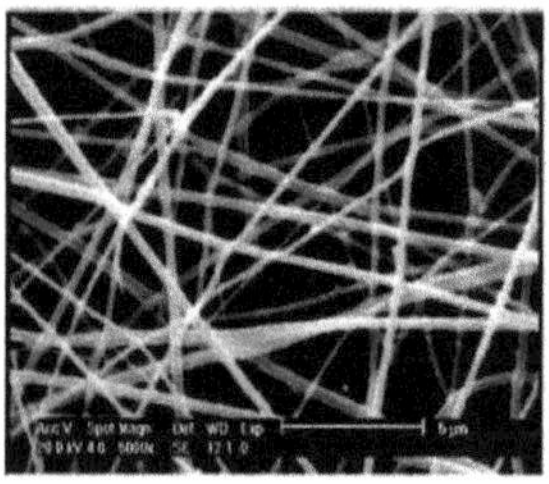	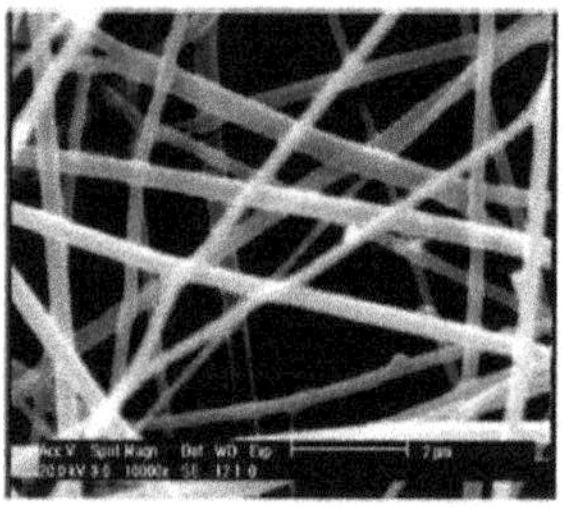	Max: 352 Min: 110 Avg: 244

FIGURE 10.6   (*Continued*)

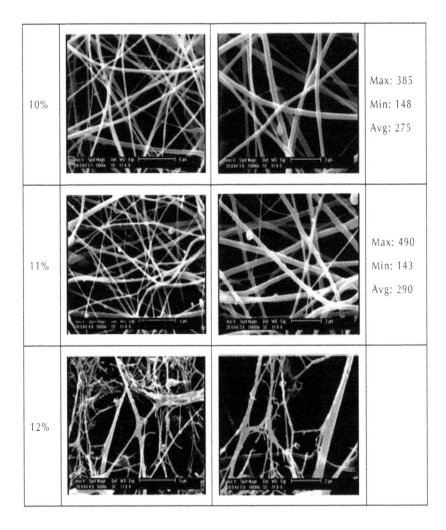

FIGURE 10.6 Scanning electron micrographs of electrospun nanofibers at different CHT concentration (wt.%): (a) 8, (b) 9, (c) 10, (d) 11, (e) 12, 24kV, 5cm, TFA/DCM: 70/30, (0.06 wt.% MWNTs).

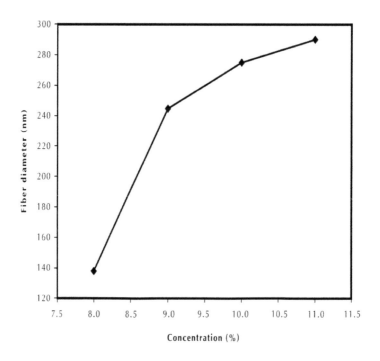

FIGURE 10.7 The effect of the CHT concentration in CHT/MWNT dispersion on nanofiber diameter.

Figure 10.7 shows the effect of the CHT concentration in CHT/MWNT dispersion on nanofiber diameter. Figure 10.8 shows the SEM image of CHT/MWNTs electrospun nanofibers produced in different voltage. In our experiments, 18kV attains as threshold voltage, where fiber formation occurs.

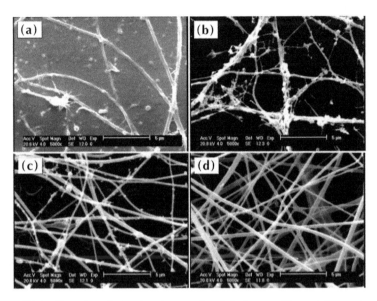

FIGURE 10.8 Scanning electronic micrographs of electrospun fibers at different voltage (kV): (a) 18, (b) 20, (c) 22, (d) 24, 5cm, 10 wt.%, TFA/DCM: 70/30. (0.06 wt.% MWNTs).

For lower voltage, the beads and some little fiber deposited on collector (Figure 10.8a). As shown in Figure 10.8a–d, the beads decreases while voltage increasing from 18 to 24kV. The collected nanofibers by applying 18kV (10.9a) and 20kV (10.9b) were not quite clear and uniform. The higher the applied voltage, the uniform nanofibers with narrow distribution starts to form. The average diameter of fibers, 22kV (10.9c), and 24kV (10.9d), respectively, were 204 (79–391), and 275 (148–385). The conductivity measurement given in Table 10.2 confirms our observation in first set of conductivity data. As can be seen from last row, the amount of electrical conductivity reaches to a maximum level of 9×10^{-5}S/cm at prescribed setup.

The distance between tips to collector is another parameter that controls the fiber diameter and morphology. Figure 10.9 shows the change in morphologies of CHT/MWNTs electrospun nanofibers at different distance. When the distance is not long enough, the solvent could not find opportunity for separation. Hence, the interconnected thick nanofiber deposits on the collector (Figure 10.9a). However, the adjusting of the distance on 5cm (Figure 10.9b) leads to homogenous nanofibers with negligible beads and interconnected areas. However, the beads increases by increasing of distance of the tip-to-collector as represented in Figure 10.9b–f. Also, the results show that the diameter of electrospun fibers decreases by increasing of distance tip to collector in Figure 10.9b, c, and d, respectively, 275 (148–385), 170 (98–283), 132 (71–224). A remarkable defects and non homogeneity appears for those fibers prepared at a distance of 8cm (Figure 10.9e) and 10cm (Figure 10.10f). However, a 5cm distance selected as proper amount for CHT/MWNTs electrospinning process.

The non homogeneity and huge bead densities plays as a barrier against electrical current and still a bead free and thin nanofiber mat shows higher conductivity compared to other samples. Experimental framework in this study was based on parameter adjusting for electrospinning of conductive CHT/MWNTs nanofiber. It can be expected that, the addition of nanotubes can boost conductivity and change morphological aspects, which is extremely important for biomedical applications.

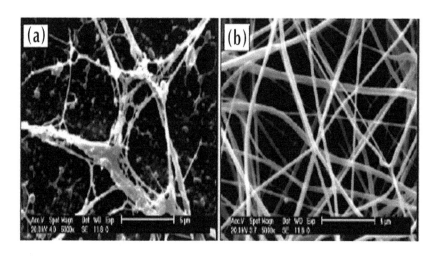

FIGURE 10.9   (*Continued*)

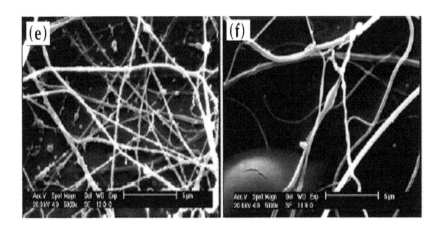

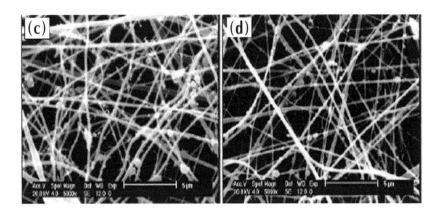

FIGURE 10.9 Scanning electronic micrographs of electrospun fibers of CHT/MWNT at different tip-to-collector distances (cm): (a) 4, (b) 5, (c) 6, (d) 7, (e) 8, (f) 10, 24kV, 10 wt.%, TFA/DCM: 70/30.

10.7 SILK FIBERS

In recent years, the electrospinning process has gained much attention be-cause it is an effective method to manufacture ultrafine fibers or fibrous structures of many polymers with diameter in the range from several micrometers down to tens of nanometers. In the electrospinning process, a high voltage is used to create an electrically charged jet of a polymer solution or a molten polymer. This jet is collected on a target as a non woven fabric. The jet typically develops a bending instability and then solidifies to form fibers, which measures in the range of nanometers to 1mm. Because these nanofibers have some useful properties such as high specific surface area and high porosity, they can be used as filters, wound dressings, tissue engineering scaffolds, and so on.

Recently, the protein-based materials have been interested in biomedical and biotechnological fields. The silk polymer, a representative fibrous protein, has been investigated as one of promising resources of biotechnology and biomedical materials due to its unique properties. Furthermore, high molecular weight synthetic polypeptides of precisely controlled amino acid composition and sequence have been fabricated using the recombinant tools of molecular biology. These genetic engineered polypeptides have attracted the researcher's attention as a new functional material for biotechnological applications including cellular adhesion promoters, biosensors, and suture materials.

Silks are generally defined as fibrous proteins that are spun into fibers by some Lepidoptera larvae such as silkworms, spiders, scorpions, mites and flies. Silkworm silk has been used as a luxury textile material since 3000 B.C., but it was not until recently that the scientific community realized the tremendous potential of silk as a structural material. While stiff and strong fibers (such as carbon, aramid, glass, etc.) are routinely manufactured nowadays, silk fibers offer a unique combination of strength and ductility which is unrivalled by any other natural or man-made fibers. Silk has excellent Properties such as lightweight (1.3g/cm^3) and high tensile strength (up to 4.8GPa as the strongest fiber known in nature). Silk is thermally stable up to 250°C, allowing processing over a wide range of temperatures. The origins of this behavior are to be found in the organization of the silks at the molecular and supramolecular levels, which are able to impart a large load bearing capability together with a damage tolerant response. The analysis of the fracture micro mechanisms in silk may shed light on the relationship between its microstructure and the mechanical properties. The anti-parallel β pleated structure of silk gives rise to its structural properties of combined strength and toughness.

B. mori silk consists of two types of proteins, fibroin and sericin. Fibroin is the protein that forms the filaments of silkworm silk. Fibroin filaments made up of bundles of nanofibrils with a bundle diameter of around 100nm. The nanofibrils are oriented parallel to the axis of the fiber, and are thought to interact strongly with each other. Fibroin contains 46% Glycine, 29% Alanine and 12% Serine. Fibroin is a giant molecule (4,700 amino acids) comprising a "crystalline" portion of about two-thirds and an "amorphous" region of about one-third. The crystalline portion comprises about 50 repeats of polypeptide of 59 amino acids whose sequence is known: Gly-Ala-Gly-Ala-Gly-Scr-Gly-Ala-Ala-Gly-(Scr-Gly-Ala-Gly-Ala-Gly)s-Tyr. This repeated unit forms, β—sheet and is responsible for the mechanical properties of the fiber.

Silk fibroin (SF) can be prepared in various forms such as gels, powders, fibers, and membranes. Number of researchers has investigated silk-based nanofibers as one of the candidate materials for biomedical applications, because it has several distinctive biological properties including good biocompatibility, good oxygen and water vapor permeability, biodegradability, and minimal inflammatory reaction. Several researchers have also studied processing parameters and morphology of electrospun silk nanofibers using hexafluoroacetone, hexafluoro-2-propanol and formic acid as solvents. In all these reports nanofibers with circular cross sections have been observed.

In this section, effects of electrospinning parameters are studied and nanofibers dimensions and morphology are reported. Morphology of fibers diameter of silk precursor were investigated varying concentration, temperature and applied voltage and observation of ribbon like silk nanofibers were reported. Furthermore, a more systematic understanding of the process conditions was studied and a quantitative basis for the relationships between average fiber diameter and electrospinning parameters was established using response surface methodology (RSM), which will provide some basis for the preparation of silk nanofibers with desired properties.

10.7.1   PREPARATION OF REGENERATED SF SOLUTION

Raw silk fibers (B. mori cocoons were obtained from domestic producer, Abrisham Guilan Co., IRAN) were degummed with 2gr/L Na_2CO_3 solution and 10gr/L anionic detergent at 100°C for 1hr and then rinsed with warm distilled water. Degummed silk (SF) was dissolved in a ternary solvent system of $CaCl_2/CH_3CH_2OH/H_2O$ (1:2:8 in molar ratio) at 70°C for 6hr. After dialysis with cellulose tubular membrane (Bialysis Tubing D9527 Sigma) in H_2O for 3 days, the SF solution was filtered and lyophilized to obtain the regenerated SF sponges.

10.7.2   PREPARATION OF THE SPINNING SOLUTION

Silk fibroin solutions were prepared by dissolving the regenerated SF sponges in 98% formic acid for 30min. Concentrations of SF solutions for electrospinning was in the range from 8 to 14% by weight.

10.7.3   ELECTROSPINNING

In the electrospinning process, a high electric potential (Gamma High voltage) was applied to a droplet of SF solution at the tip (0.35mm inner diameter) of a syringe needle, The electrospun nanofibers were collected on a target plate which was placed at a distance of 10cm from the syringe tip. The syringe tip and the target plate were enclosed in a chamber for adjusting and controlling the temperature. Schematic diagram of the electrospinning apparatus is shown in Figure 10.10. The processing temperature was adjusted at 25, 50 and 75°C. A high voltage in the range from 10 to 20kV was applied to the droplet of SF solution.

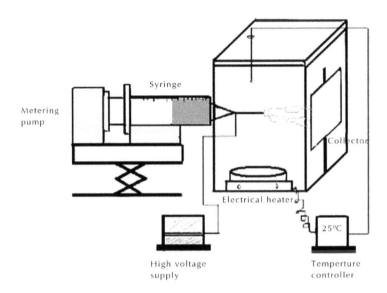

FIGURE 10.10   Schematic diagram of electrospinning apparatus.

10.7.4   CHARACTERIZATION

Optical microscope (Nikon Microphot-FXA) was used to investigate the macroscopic morphology of electrospun SF fibers. For better resolving power, morphology, surface texture and dimensions of the gold-sputtered electrospun nanofibers were determined using a Philips XL-30 scanning electron microscope. A measurement of about 100 random fibers was used to determine average fiber diameter and their distribution.

10.7.5   RESULTS AND DISCUSSION

EFFECT OF SILK CONCENTRATION

One of the most important quantities related with electrospun nanofibers is their diameter. Since nanofibers are resulted from evaporation of polymer jets, the fiber diameters will depend on the jet sizes and the solution concentration. It has been reported that during the traveling of a polymer jet from the syringe tip to the collector, the primary jet may be split into different sizes multiple jets, resulting in different fiber diameters. When no splitting is involved in electrospinning, one of the most important parameters influencing the fiber diameter is Concentration of regenerated silk solution. The jet with a low concentration breaks into droplets readily and a mixture of fibers, bead fibers and droplets as a result of low viscosity is generated. These fibers have an irregular morphology with large variation in size, on the other hand jet with high concentration don't break up but traveled to the grounded target and tend to facilitate the formation of fibers without beads and droplets. In this case, Fibers became more uniform with regular morphology.

At first, a series of experiments were carried out when the silk concentration was varied from 8 to 14% at the 15KV constant electric field and 25°C constant temperature. Below the silk concentration of 8% as well as at low electric filed in the case of 8% solution, droplets were formed instead of fibers. Figure 10.11 shows morphology of the obtained fibers from 8% silk solution at 20KV. The obtained fibers are not uniform. The average fiber diameter is 72nm and a narrow distribution of fiber diameters is observed. It was found that continues nanofibers were formed above silk concentration of 8% regardless of the applied electric field and electrospinning condition.

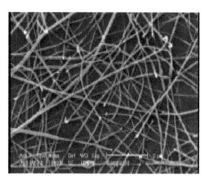 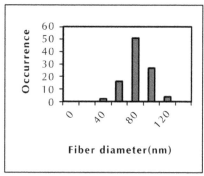

FIGURE 10.11   SEM micrograph and fiber distribution of 8 wt.% of silk at 20KV and 25°C.

In the electrospinning of silk fibroin, when the silk concentration is more than 10%, thin and rod like fibers with diameters range from 60–450nm were obtained. Figures 10.12–10.14 show the SEM micrographs and diameter distribution of the resulted fibers

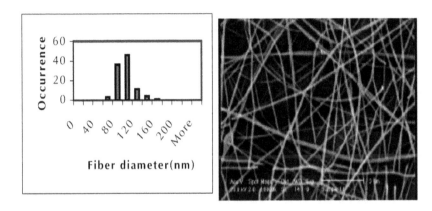

FIGURE 10.12   SEM micrograph and fiber distribution of 10 wt.% of silk at 15KV and 25°C.

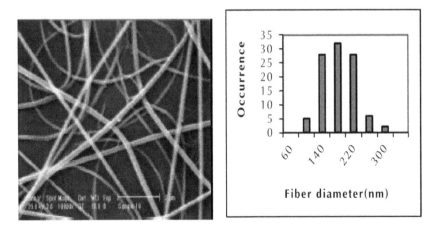

FIGURE 10.13   SEM micrograph and fiber distribution of 12 wt.% of silk at 15KV and 25°C.

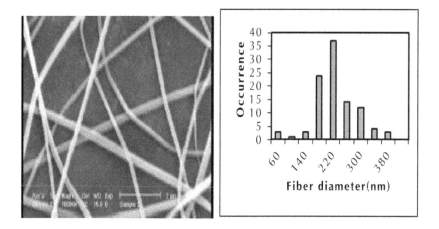

FIGURE 10.14 SEM micrograph and fiber distribution of 14 wt.% of silk at 15KV and 25°C.

There is a significant increase in mean fiber diameter with the increasing of the silk concentration, which shows the important role of silk concentration in fiber formation during electrospinning process. Concentration of the polymer solution reflects the number of entanglements of polymer chains in the solution, thus solution viscosity. Experimental observations in electrospinning confirm that for forming fibers, a minimum polymer concentration is required. Below this critical concentration, application of electric field to a polymer solution results electrospraying and formation of droplets to the instability of the ejected jet. As the polymer concentration increased, a mixture of beads and fibers is formed. Further increase in concentration results in formation of continuous fibers as reported in this paper. It seems that the critical concentration of the silk solution in formic acid for the formation of continuous silk fibers is 10%.

Experimental results in electrospinning showed that with increasing the temperature of electrospinning process, concentration of polymer solution has the same effect on fibers diameter at 25°C. Figures 10.15–10.17 show the SEM micrographs and diameter distribution of the fibers at 50°C.

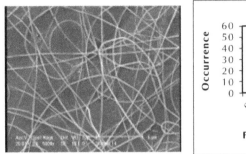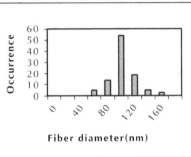

FIGURE 10.15 SEM micrograph and fiber distribution of 10 wt.% of silk at 15KV and 50°C.

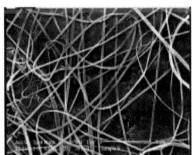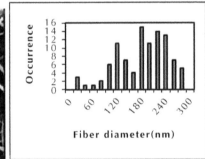

FIGURE 10.16 SEM micrograph and fiber distribution of 12 wt.% of silk at 15KV and 50°C.

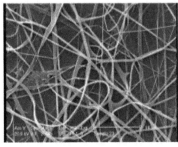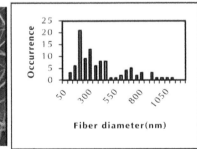

FIGURE 10.17 SEM micrograph and fiber distribution of 14 wt.% of silk at 15KV and 50°C.

When the temperature of the electrospinning process increased, the circular cross section became elliptical and then flat, forming a ribbon with a cross-sectional perimeter nearly the same as the perimeter of the jet. Flat and ribbon like fibers have greater diameter than circular fibers. It seems that at high voltage the primary jet splits into different sizes multiple jets, resulting in different fiber diameters that nearly in a same range of diameter.

At 75°C, the concentration and applied voltage have same effect as 50°C. Figures 10.18–10.20 show the SEM micrographs and diameter distribution of nano fibers at 75°C.

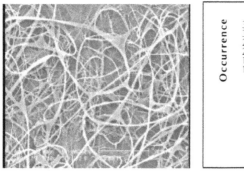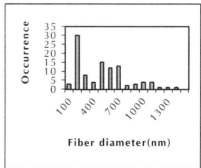

FIGURE 10.18   SEM micrograph and fiber distribution of 10 wt.% of silk at 15KV and 75°C.

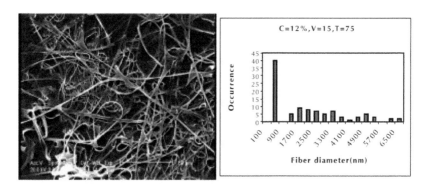

FIGURE 10.19   SEM micrograph and fiber distribution of 12 wt.% of silk at 15KV and 75°C.

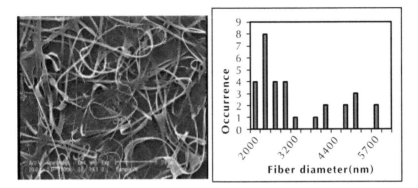

FIGURE 10.20   SEM micrograph and fiber distribution of 14 wt.% of silk at 15KV and 75°C.

Figure 10.21 shows the relationship between mean fiber diameter and SF concentration at different electrospinning temperature. There is a significant increase in mean fiber diameter with increasing of the silk concentration, which shows the important role of silk concentration in fiber formation during electrospinning process. It is well known that the viscosity of polymer solutions is proportional to concentration and polymer molecular weight. For concentrated polymer solution, concentration of the polymer solution reflects the number of entanglements of polymer chains, thus have considerable effects on the solution viscosity. At fixed polymer molecular weight, the higher polymer concentration resulting higher solution viscosity. The jet from low viscosity liquids breaks up into droplets more readily and few fibers are formed, while at high viscosity, electrospinning is prohibit because of the instability flow causes by the high cohesiveness of the solution. Experimental observations in electrospinning confirm that for fiber formation to occur, a minimum polymer concentration is required. Below this critical concentration, application of electric field to a polymer solution results electro spraying and formation of droplets to the instability of the ejected jet. As the polymer concentration increased, a mixture of beads and fibers is formed. Further increase in concentration results in formation of continuous fibers as reported in this chapter. It seems that the critical concentration of the silk solution in formic acid for the formation of continuous silk fibers is 10% when the applied electric field was in the range of 10 to 20kV.

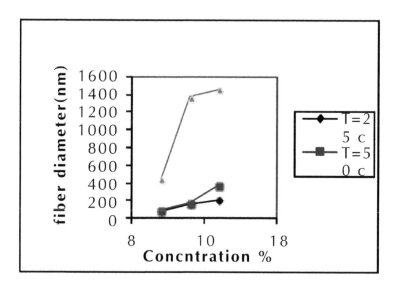

FIGURE 10.21   Mean fiber diameter of electrospun silk fibers at 25, 50 and 75°C at 15kV.

EFFECT OF ELECTRIC FIELD

It was already reported that the effect of the applied electrospinning voltage is much lower than effect of the solution concentration on the diameter of electrospun fibers. In order to study the effect of the electric field, silk solution with the concentration of 10, 12, and 14% were electrospun at 10, 15, and 20KV at 25°C. The variation of the mean fiber diameter at different applied voltage for different concentration is shown in Figure 10.22 This figure and SEM micrographs show that at all electric fields continuous and uniform fibers with different fiber diameter are formed. At a high solution concentration, Effect of applied voltage is nearly significant. It is suggested that, at this temperature, higher applied voltage causes multiple jets formation, which would provide decrees fiber diameter.

As the results of this finding it seems that electric field shows different effects on the nanofibers morphology. This effect depends on the polymer solution concentration and electrospinning conditions.

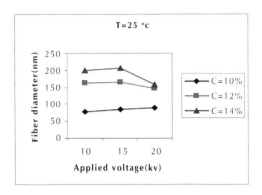

FIGURE 10.22   Mean fiber diameter of electrospun silk fibers at 10, 15, 20 and 25°C.

EFFECT OF ELECTROSPINNING TEMPERATURE

One of the most important quantities related with electrospun nanofibers is their diameter. Since nanofibers are resulted from evaporation of polymer jets, the fiber diameters will depend on the jet sizes. The elongation of the jet and the evaporation of the solvent both change the shape and the charge per unit area carried by the jet. After the skin is formed, the solvent inside the jet escapes and the atmospheric pressure tends to collapse the tube like jet. The circular cross section becomes elliptical and then flat, forming a ribbon-like structure. In this work we believe that ribbon-like structure in the electrospinning of SF at higher temperature thought to be related with skin formation at the jets. With increasing the electrospinning tempera-ture, solvent evaporation rate increases, which results in the formation of skin at the jet surface. Non uniform lateral stresses around the fiber due to the uneven evaporation of solvent and/or striking the target make the nanofibers with circular cross-section to collapse into ribbon shape.

Bending of the electrospun ribbons were observed on the SEM micro-graphs as a result of the electrically driven bending instability or forces that occurred when the ribbon was stopped on the collector. Another prob-lem that may be occurring in the electrospinning of SF at high temperature is the branching of jets. With increasing the temperature of electrospinning process, the balance between the surface tension and electrical forces can shift so that the shape of a jet becomes unstable. Such an unstable jet can reduce its local charge per unit surface area by ejecting a smaller jet from the surface of the primary jet or by splitting apart into two smaller jets. Branched jets, resulting from the ejection of the smaller jet on the surface of the primary jet were observed in electrospun fibers of SF. The axes of the cones from which the secondary jets originated were at an angle near 90° with respect to the axis of the primary jet. Figure 10.23 shows the SEM micrographs of flat, ribbon like and branched fibers.

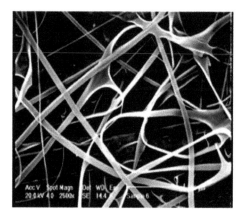

FIGURE 10.23   SEM micrograph of silk flat, ribbon like and branched nano fibers at high temperature.

In order to study the effect of electrospinning temperature on the morphology and texture of electrospun silk nanofibers, 12% silk solution was electrospun at various temperatures of 25, 50 and 75°C. Results are shown in Figure 10.24 Interestingly, the electrospinning of silk solution showed flat fiber morphology at 50 and 75°C, whereas circular structure was observed at 25°C. At 25°C, the nanofibers with a rounded cross section and a smooth surface were collected on the target. Their diameter showed a size range of approximately 100–300nm with 180nm being the most frequently occurring. They are within the same range of reported size for electrospun silk nanofibers. With increasing the electrospinning temperature to 50°C, The morphology of the fibers was slightly changed from circular cross section to ribbon like fibers. Fiber diameter was also increased to a range of approximately 20–320nm with 180nm the most occurring frequency. At 75°C, The morphology of the fibers was completely changed to ribbon like structure. Furthermore, fibers dimensions were increased significantly to the range of 500–4,100nm with 1,100nm the most occurring frequency.

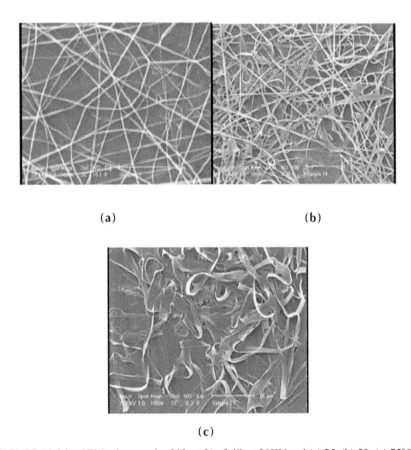

(a)   (b)

(c)

FIGURE 10.24   SEM micrograph of 12 wt.% of silk at 20KV and (a) 25, (b) 50, (c) 75°C.

10.8 MODELING

Response surface methodology (RSM) is a collection of mathematical and statistical techniques for empirical model building (Appendix). By careful design of *experiments*, the objective is to optimize a *response* (output variable) which is influenced by several *independent variables* (input variables). An experiment is a series of tests, called *runs*, in which changes are made in the input variables in order to identify the reasons for changes in the output response.

In order to optimize and predict the morphology and average fiber diameter of electrospun silk, design of experiment was employed in the present work. Morphology of fibers and distribution of fiber diameter of silk precursor were investigated varying concentration, temperature and applied voltage. A more systematic understanding of these process conditions was obtained and a quantitative basis for the relationships between average fiber diameter and electrospinning parameters was established using response surface methodology (Appendix), which will provide some basis for the preparation of silk nanofibers.

A central composite design was employed to fit a second-order model for three variables. Silk concentration (X_1), applied voltage (X_2), and temperature (X_3) were three independent variables (factors) considered in the preparation of silk nanofibers, while the fibers diameter were dependent variables (response). The actual and corresponding coded values of three factors (X_1, X_2, and X_3) are given in Table 10.1. The following second-order model in X_1, X_2 and X_3 was fitted using the data in Table 10.1:

$$Y = \beta_0 + \beta_1 x_1 + \beta_2 x_2 + \beta_3 x_3 + \beta_1 x_1^2 + \beta_2 x_2^2 + \beta_3 x_3^2 + \beta_1 x_1 x_2 + \beta_1 x_1 x_3 + \beta_2 x_2 x_3 + \varepsilon$$

TABLE 10.1   Central composite design

		Coded values		
X_i	Independent variables	-1	0	1
X_i	Silk concentration (%)	10	12	14
X_i	applied voltage (KV)	10	15	20
X_i	temperature (° C)	25	50	75

The Minitab and Mathlab programs were used for analysis of this second-order model and for response surface plots (Minitab 11, Mathlab 7).

By Regression analysis, values for coefficients for parameters and P-values (a measure of the statistical significance) are calculated. When P-value is less than 0.05, the factor has significant impact on the average fiber diameter. If P-value is greater than 0.05, the factor has no significant impact on average fiber diameter. And R^2_{adj} (represents the proportion of the total variability that has been explained by the regression model) for regression models were obtained (Table 10.2) and main effect plots on fiber diameter (Figure 10.25) were obtained and reported.

The fitted second-order equation for average fiber diameter is given by:

$Y = 391 + 311\ X_1 - 164\ X_2 + 57\ X_3 - 162\ X_1^2 + 69\ X_2^2 + 391\ X_3^2 - 159\ X_1X_2 + 315\ X_1X_3 - 144\ X_2X_3$ (10.2)

Where Y = Average fiber diameter

TABLE 10.2 Regression Analysis for the three factors (concentration, applied voltage, temperature) and coefficients of the model in coded unit*

Variables	Constant		P-value
	β_0	391.3	0.008
x1	β_1	310.98	0.00
x2	β_2	-164.0	0.015
x3	β_3	57.03	0.00
x21	β_{11}	161.8	0.143
x22	β_{22}	68.8	0.516
x23	β_{33}	390.9	0.002
x1x2	β_{12}	-158.77	0.048
x1x3	β_{13}	314.59	0.001
x2x3	β_{23}	-144.41	0.069
F	P-value	R2	R2 (adj)
18.84	0.00	0.907	0.858

*Model: $Y = \beta_0 + \beta_1 x_1 + \beta_2 x_2 + \beta_3 x_3 + \beta_{11} x_1^2 + \beta_{22} x_2^2 + \beta_{332} x_3^2 + \beta_{12} x_1 x2 + \beta_{13} x_1 x\ 3 + \beta_{13} x_2 x3$ where "y" is average fiber diameter.

From the P-values listed in Table 10.2, it is obvious that P-value of term X_2 is greater than P-values for terms X_1 and X_3. And other P-values for terms related to applied voltage such as, X_2^2, X_1X_2, X_2X_3 are much greater than significance level of 0.05. That is to say, applied voltage has no much significant impact on average fiber diameter and the interactions between concentration and applied voltage, temperature and applied voltage are not significant, either. But P-values for term related to X_3 and X_1 are less than 0.05. Therefore, temperature and concentration have significant impact on average fiber diameter. Furthermore, R^2_{adj} is 0.858, That is to say, this model explains 86% of the variability in new data.

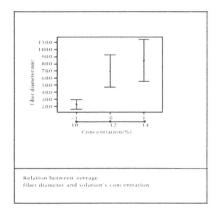

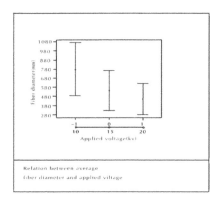

FIGURE 10.25   *Continued*

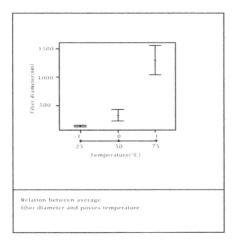

FIGURE 10.25   Effect of electrospinning parameters on silk nano fibers diameter.

10.9   CONCUDING REMARKS

Several solvents, including acetic acid 1–90%, formic acid and TFA/DCM (70:30), were used for electrospinning of chitosan/carbon nanotube dispersion. It has been observed that the TFA/DCM (70:30) solvent is the only solvent with a proper reliablity for the electrospinnability of chitosan/ carbon nanotube. This is a significant improvement in the electrospinning of chitosan/carbon nanotube dispersion. It was also observed that homogenous fibers with an average diameter of 455nm (306–672) could be prepared with chitosan/carbon nanotube dispersion in TFA/DCM 70:30. In addition, the SEM images showed that the fiber diameter decreased by decreasing voltage and increasing the tip-to-collector distance.

The electrospinning of silk fibroin was processed and the average fiber diameters depend on the electrospinning condition. Morphology of fibers and distribution of diameter were investigated at various concentrations, applied voltages and temperature. The electrospinning temperature and the solution concentration have a significant effect on the morphology of the electrospun silk nanofibers. There effects were explained to be due to the change in the rate of skin formation and the evaporation rate of solvents. To determine the exact mechanism of the conversion of polymer into nanofibers require further theoretical and experimental work.

From the practical view the results of the present work can be condensed. Concentration of regenerated silk solution was the most dominant parameter to produce uniform and continuous fibers. The jet with a low concentration breaks into droplets readily and a mixture of fibers and droplets as a result of low viscosity is generated. On the other hand jets with high concentration do not break up but traveled to the target and tend to facilitate the formation of fibers without beads and droplets. In this case, fibers become more uniform with regular morphology. In the electrospinning of silk fibroin, when the silk concentration is more than 10%, thin and rod like fibers with diameters range from 60–450nm were obtained. Furthermore, In the electrospinning of silk fibroin, when the process temperature is more than 25°C, flat, ribbon like and branched fibers with diameters range from 60 to 7,000nm were obtained.

Two-way analysis of variance was carried out at the significant level of 0.05 to study the impact of concentration, applied voltages and temperature on average fiber diameter. It was concluded that concentration of solution and electrospinning temperature were the most significant factors impacting the diameter of fibers. Applied voltage had no significant impact on average fiber diameter. The average fiber diameter increased with polymer concentration and electrospinning temperature according to proposed relationship under the experimental conditions studied in this chapter.

APPENDIX-A

Materials produced from fibers and threads are classified as textile: fabrics, nonwoven materials, fur fabric, carpets and rugs, and so on.

Textile fibers are the main raw material of the textile industry. According to their origin, these fibers are divided into natural and chemical ones.

Natural fibers are plant (cotton, bast fibers), animal (wool, silk) and mineral (asbestos) origin ones.

Chemical fibers are produced from modified natural or synthetic high-molecular substances and are classified as artificial ones obtained by chemical processing of natural raw material, commonly cellulose (viscose, acetate), and synthetic ones obtained from synthetic polymers (nylon-6, polyester, acryl, PVC fibers, etc.).

Textile materials are damaged by microorganisms, insects, rodents and other biodamaging agents. Fibers and fabrics resistance to biodamages primarily depends upon chemical nature of the fibers from which they are made. Most frequently we have to put up with microbiological damages of textile materials based on natural fibers—cotton, linen, and so on, utilized by saprophyte microflora. Chemical fibers and fabrics, especially synthetic ones, are higher biologically resistant, but microorganisms-biodegraders can also adapt to them.

Textile material degradation by microorganisms depends on their wear rate, kind and origin, organic composition, temperature and humidity conditions, degree of aeration, and so on.

With increased humidity and temperature, and restricted air exchange microorganisms damage fibers and fabrics at different stages of their manufacture and application, starting from the primary processing of fibers including spinning, weaving, finishing and storage, transportation and operation of textile materials and articles from them. Fiber and fabric biodamage intensity sharply increases when contacting with the soil and water, specifically in the regions with warm and humid climate.

Textile materials are damaged by bacteria and microscopic fungi. Bacterial degradation of textile materials is more intensive than the fungal one. The damaging bacterium genuses are: *Cytophaga, Micrococcus, Bacterium, Bacillus, Cellulobacillus, Pseudomonas, Sarcina*. Among fungi damaging textile materials in the air and in the soil the following are detected: *Aspergillus, Penicillium, Alternaria, Cladosporium, Fusarium, Trichoderma*, and so on.

Annual losses due to microbiological damaging of fabrics reach hundreds of millions of dollars.

Fiber and fabric biodamaging by microorganisms is usually accompanied by the mass loss and mechanical strength of the material as a result, for example, of fiber degradation by microorganism metabolites: enzymes, organic acids, and so on.

CURIOUS FACTS

The immediate future of the textile industry belongs to biotechnology. Even today suggestions on the synthesis of various polysaccharides using microbiological methods are present. These methods may be applied to synthesis of fiber-forming monomers and polymers.

Scientists have demonstrated possibilities of microbiological synthesis of some monomers to produce dicarboxylic acids, caprolactam, and so on. Some kinds of fiber-forming polymers, polyethers, in particular, can also be obtained by microbiological synthesis.

Some kinds of fiber-forming polypeptides have already been obtained by the microbiological synthesis. In some cases, concentration of these products may reach 40% of the biomass weight and they can be used as a perspective raw material for synthetic fibers. Studies in this direction are widely performed in many countries all over the world.

The impact of microorganisms on textile materials that causes their degradation is performed, at least, by two main ways (direct and indirect):
- Fungi and bacteria use textile materials as the nutrient source (assimilation);
- Textile materials are damaged by microorganism metabolism (degradation).

Biodamages of textile materials induced by microorganisms and their metabolites manifest in coloring (occurrence of spots on textile materials or their coatings), defects (formation of bubbles on colored surfaces of textile materials), bond breaks in fibrous materials, penetration deep inside (penetration of microorganisms into the cavity of the natural fiber), deterioration of mechanical properties (e.g. strength at break reduction), mass loss, a change of chemical properties (cellulose degradation by microorganisms), liberation of volatile substances and changes of other properties.

It is known that when microorganisms completely consume one part of the substrate they then are able to liberate enzymes degrading other components of the culture medium. It is found that, degrading fiber components each group of microorganisms due to their physiological features decomposes some definite part of the fiber, damages it differently and in a different degree. It is found that along with the enzymes, textile materials are also degraded by organic acids produced by microorganisms: lactic, gluconic, acetic, succinic, fumaric, malic, citric, oxalic, and so on. It is also found that enzymes and organic acids liberated by microorganisms continue degrading textile materials even after microorganisms die. As noted, the content of cellulose, proteins, pectins and alcohol-soluble waxes increases with the fiber damage degree in it; pH increases, and concentration of water-soluble substances increases that, probably, is explained by increased accumulation of metabolites and consumption of nutrients by microorganisms for their vital activity.

The typical feature of textile material damaging by microorganisms is occurrence of honey dew, red violet or olive spots with respect to a pigment produced by microorganisms and fabric color. As microorganism's pigment interacts with the fabric dye, spots of different hue and tints not removed by laundering or by hydrogen peroxide oxidation. They may sometimes be removed by hot treatment in blankly solution. Spot occurrence on textile materials is usually accompanied by a strong musty odor.

CURIOUS FACTS

Textile materials are also furnished by biotechnological methods based on the use of enzymes performing various physicochemical processes. Biotechnologies are mostly full for preparation operations (cloth softening, boil-off and bleaching of cotton fabrics, wool washing) and bleaching effects of jeans and other fabrics, biopolishing, and making articles softer.

Enzymes are used to improve sorption properties of cellulose fibers, to increase specific area and volume of fibers, to remove pectin "companions" of cotton and linen cellulose. Enzymes also hydrolyze ether bonds on the surface of polyether fibers.

Cotton fabric dyeing technologies in the presence of enzymes improving coloristic parameters of prepared textile articles under softer conditions, which reduce pollutants in the sewage. To complete treatment of the fabric surface, that is removal of surface fiber fibrilla, enzymes are applied. The enzymatic processing also decreases and even eliminates the adverse effect of long wool fiber puncturing. The application of enzymes to treatment of dyed tissues is industrially proven. This processing causes irregular bleaching that adheres the articles fashionable "worn" style.

Temperature and humidity are conditions promoting biodamaging of fibrous materials. The comparison of requirements to biological resistance of textile materials shall be based on the features of every type of fibers, among which mineral fiber are most biologically resistant.

Different microorganism damaging rate for fabrics is due to their different structure. Thinner fabrics with the lower surface density and higher through porosity are subject to the greatest biodamaging, because these properties of the material provide large contact area for microorganisms and allow their easy penetration deep into it. The thread bioresistance increases with the yarning rate.

Cotton fiber is a valuable raw material for the textile industry. Its technological value is due to a complex of properties that should be retained during harvesting, storage, primary and further processing to provide high quality of products. One of the factors providing retention of the primary fiber properties is its resistance to bacteria and fungi impacts. This is tightly associated with the chemical and physical features of cotton fiber structures.

Ripe cotton fiber represents a unit extended plant cell shaped as a flattened tube with a corkscrew waviness. The upper fiber end is cone-shaped and dead-ended. The lower end attached to the seed is a torn open channel.

The cotton fiber structure is formed during its maturation, when cellulose is biosynthesized and its macromolecules are regularly disposed.

The basic elements of cotton fiber morphological structure are known to be the cuticle, primary wall, convoluted layer, secondary wall, and tertiary wall with the central channel. The fiber surface is covered by a thin layer of wax substances—the primary wall (cuticle). This layer is a protective one and possesses rather high chemical resistance.

There is a primary wall under the cuticle consisting of a cellulose framework and fatty-wax-pectin substances. The upper layer of the primary wall is less densely packed as compared with the inner one, due to fiber surface expansion during its growth. Cellulose fibrils in the primary wall are not regularly oriented.

The primary layer covers the convoluted layer having structure different from the secondary wall layers. It is more resistant to dissolution, as compared with the main cellulose mass.

The secondary cotton fabric wall is more homogeneous and contains the greatest amount of cellulose. It consists of densely packed, concurrently oriented cellulose fibrils composed in thin layers. The fibril layers are spiral-shaped twisted around the fiber axis. There are just few micropores in this layer.

The tertiary wall is an area adjacent to the fiber channel. Some authors think that the tertiary wall contains many pores and consists of weakly ordered cellulose fibrils and plenty of protein admixtures, protoplasm, and pectin substances.

The fiber channel is filled with protoplasm residues which are proteins, and contains various mineral salts and a complex of microelements. For the mature fiber, channel cross-section is 4–8% of total cross-section.

Cotton fiber is formed during maturation. This includes not only cellulose biosynthesis, but also ordering of the cellulose macromolecules shaped as chains and formed by repetitive units consisting two β-D-glucose residues bound by glucosidic bonds.

Among all plant fibers, cotton contains the maximum amount of cellulose (95–96%).

The morphological structural unit of cellulose is a cluster of macromolecules—a fibril 1.0–1.5μm long and 8–15nm thick, rather than an individual molecule. Cellulose fibers consist of fibril clusters uniform oriented along the fiber or at some angle to it.

It is known that cellulose consists not only of crystalline areas—micelles where molecule chains are concurrently oriented and bound by intermolecular forces, but also of amorphous areas.

Amorphous areas in the cellulose fiber are responsible for the finest capillaries formation, that is a "sub-microscopic" space inside the cellulose structure is formed. The presence of the sub-microscopic system of capillaries in the cellulose fibers is of paramount significance, because it is the channel of chemical reactions by which water-soluble reagents penetrate deep in the cellulose structure. More active hydroxyl groups interacting with various substances are also disposed here.

It is known that hydrogen bonds are present between hydroxyl groups of cellulose molecules in the crystalline areas. Hydroxyl groups of the amorphous area may occur free of weakly bound and, as a consequence, they are accessible for sorption. These hydroxyl groups represent active sorption centers able to attract water.

Among all plant fibers, cotton has the highest quantity of cellulose (95–96%). Along with cellulose, the fibers contain some fatty, wax, coloring mineral substances (4–5%). Cellulose concomitant substances are disposed between macromolecule clusters and fibrils. Raw cotton contains mineral substances (K, Na, Ca, Mg) that promote mold growth and also contains microelements (Fe, Cu, Zn) stimulating growth of microorganisms. Moreover, it contains sulfates, phosphorus, glucose, glycidols and nitrogenous substances, which also stimulate growth of microbes. Differences in their concentrations are one of the reasons for different aggressiveness of microorganisms in relation to the cotton fiber.

The presence of cellulose, pectin, nitrogen-containing and other organic substances in the cotton fiber, as well as its hygroscopicity makes it a good culture medium for abundant microflora.

Cotton is infected by microorganisms during harvesting, transportation and storage. When machine harvested, the raw cotton is clogged by various admixtures. It obtains multiple fractures of leaves and cotton-seed hulls with humidity higher than of the fibers. Such admixtures create a humid macrozone around, where microorganisms intensively propagate. Fiber humidity above 9% is the favorable condition for cotton fiber degradation by microorganisms.

It is found that cotton fiber damage rate directly in hulls may reach 42–59%; hence, the fiber damage rate depends on a number of factors, for example cultivation conditions, harvesting period, type of selection, and so on.

Cotton fiber maturity is characterized by filling with cellulose. As the fiber becomes more mature, its strength, elasticity and coloring value increase.

Low quality cotton having higher humidity is damaged by microorganisms to a greater extent. The fifths fibers contain microorganisms 3–5 times more than the first quality fibers. When cotton hulls open, the quantity of microorganisms sharply increases in them, because along with dust wind brings fungus spores and bacteria to the fibers.

Cotton is most seriously damaged during storage: in compartments with high humidity up to 24% of cotton is damaged. Cotton storage in compact bales covered by tarpaulin is of high danger, especially after rains. For instance, after one and half months of such storage the fiber is damaged by 50% or higher.

Cotton microflora remains active under conditions of the spinning industry. As a result, the initial damage degree of cotton significantly increases.

CURIOUS FACTS

On some textile enterprises of Ivanovskaya Oblast, sickness cases of spinning-preparatory workshops were observed. When processing biologically contaminated cotton, plenty of dust particles with microorganisms present on them are liberated to the air. This can affect the health status of the employees.

Under natural conditions, cotton products are widely used in contact with the soil (fabrics for tents) receiving damage both from the inside and the outside. The main role here is played by cellulose degrading bacteria and fungi.

It was considered over a number of years that the main role in cotton fiber damage is played by cellulose degrading microorganisms. Not denying participation of cellulose degrading bacteria and fungi in damaging of the cotton fiber, it is noted that a group of bacteria with yellow pigmented mucoid colonies representing epiphytic microflora always present on the cotton plant dominate in the process of the fiber degradation. Nonspore-forming epiphytic bacteria inhabiting in the cotton plants penetrate from their seeds into the fiber channel and begin developing there. Using chemical substances of the channel, these microorganisms then permeate into the sub-microscopic space of the tertiary wall primarily consuming pectins of the walls and proteins of the channels.

Enzymes and metabolites produced by microorganisms induce hydrolysis of cellulose macromolecules, increasing damage of internal areas of the fiber. Thus, the fiber delivered to processing factories may already be significantly damaged by microorganisms that inevitably affect production of raw yarn, fabric, and so on.

135 strains of fungi of different genus capable of damaging cotton fibers are currently determined. It is found that the population of phytopathogenic fungi is much lower than that of cellulose degraders: *Chaetomium globosum, Aspergillus flavus, Aspergillus niger, Rhizopus nigricans, Trichothecium roseum.* These species significantly deteriorate the raw cotton condition; sharply reduce spinning properties of the fiber, in particular.

It is also found that the following species of fungi are usually present on cotton fibers: *Mucor* (consumes water-soluble substances), *Aspergillus* and *Penicillium* (consumes insoluble compounds), *Chaetomium, Trichoderma*, and so on (degrade cellulose). This points to the fact that some species of mold fungi induce the real fiber decomposition that shall be distinguished from simple surface growth of microorganisms. For example, *Mucor* fungi incapable of inducing cellulose degradation may actively vegetate on the yarn finish. Along with fungi, bacteria are always present on the raw cotton, most represented by *Bacillus* and *Pseudomonas* genus species.

Figures 10.26 and 10.27 present surface micrographs of the first and fifth quality grade primary cotton fibers.

Figure 10.28, a micrograph, shows the cotton fiber surface after the impact of spontaneous microflora during 7 days. It is observed that bacterial cells are accumulated in places of fiber damage, at clearly noticeable cracks. Figure 10.4 shows the first quality cotton fiber surface after impact of *Aspergillus niger* culture during 14 days. On the fiber surface mycelium is observed. Figures 10.30 and 10.31 show photos of cotton fiber surfaces infected by *Bac. subtilis* (14-day exposure): bacterial cells on first quality fibers are separated by the surface, forming no conglomerates, whereas on the fifth quality fibers conglomerates are observed that indicates their activity (Figure 10.32).

Academician A. A. Imshenetsky has demonstrated that aerobic cellulose bacteria are able to propagate under increased humidity, whereas fungi propagate at lower humidity. Textile products are destroyed by fungi at their humidity about 10%, whereas bacteria destroy them at humidity level of, at least, 20%. As a consequence, the main attention at cotton processing to yarn should be paid to struggle against fungi, and at wet spinning and finishing fabrics and knitwear not only fungi, but mostly bacteria should be struggled against.

FIGURE 10.26   The surface of initial first quality cotton specimen (×4,500).

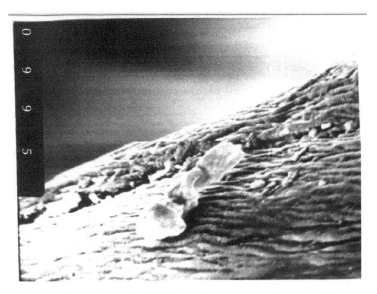

FIGURE 10.27   The surface of initial fifth quality cotton specimen (×4,500).

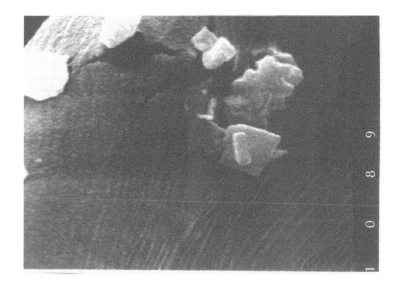

FIGURE 10.28 The first quality cotton, 7-day exposure (spontaneous microflora) (×10,000).

FIGURE 10.29 The first quality cotton, *Asp. niger* contaminated, 14-day exposure (×3,000).

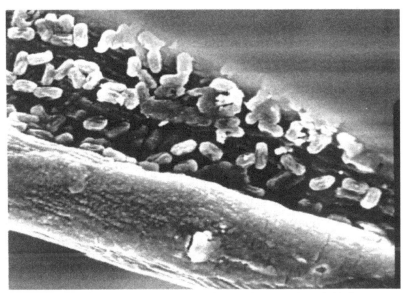

FIGURE 10.30 The first quality cotton, *Bacillus subtilis* contaminated, 14-day exposure (×4,500).

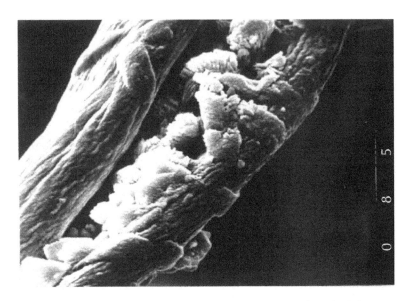

FIGURE 10.31 The fifth quality cotton, *Bacillus subtilis* contaminated, 14-day exposure (×4,500).

Cotton damage leads to:
- Significant decrease of strength of the fibers and articles from them;
- Disturbance of technological process (the smallest particles of sticky mucus excreted by some species of bacteria and fungi become the reason for sticking executive parts of machines);
- Abruptness increase;
- Waste volume increase.

Damaging of cotton fibers, fabrics and textile products by microorganisms is primarily accompanied by occurrence of colored yellow, orange, red, violet, and so on. spots and then by putrefactive odor, and, finally, the product loses strength and degrades.

The effect of microorganisms results in noticeable changes in chemical composition and physical structure of cotton fibers.

As found by electron microscopy, cotton fiber degradation by enzymes is most intensive in the zones of lower fibril structure density.

In the cotton fiber damaged by microorganisms, cellulose concentration decreases by 7.5%, pectin substances—by 60.7%, hemicellulase—by 20%, and non cellulose polysaccharides content also decreases. Cellulose biostability increases with its crystallinity degree and macromolecule orientation, as well as with hydroxyl group replacement by other functional groups. Microscopic fungi and bacteria are able to degrade cellulose and as a result glucose is accumulated in the medium, used as a source of nutrition by microorganisms. However, some part of cellulose is not destroyed and completely preserves its primary structure.

Cellulose of undamaged cotton fiber has 76.5% of well-ordered area, 7.8% of weakly ordered area, and 15.7% of disordered area. Microbiological degradation reduces the part of disordered area to 12.7%, whereas the part of well-ordered area increases to 80.4%. The ratio of weakly ordered area changes insignificantly. This goes to prove that the order degree of cotton cellulose increases due to destruction of disordered areas.

A definite type of fiber degradation corresponds to each stage of the cotton fiber damage. The initial degree of damage is manifested in streakiness, when the fiber surface obtains cracks of different length and width due to its wall break.

Swellings are formed resulting abundant accumulation of microorganisms and their metabolites in a definite part of the fiber. They may be accompanied by fiber wall break induced by biomass pressure. In this case, microorganisms and their metabolites splay out that causes blobs formation from the fiber and breaks in the yarn, as well as irregular fineness and strength.

The external microflora induces the wall damage. The highest degradation stage is fiber decomposition and breakdown into separate fibrils. Hence, perfect fiber structure is absent in this case.

In all cases of damage, a high amount of fungal mycelium may be present on the fiber surface, which hyphae penetrate through the fiber or wrap about it thus preventing spinning and coloring of textile materials.

Enzymatic activity of fungi is manifested in strictly defined places of cellulose microfibrils, and the strength loss rate depends on both external climate conditions and contamination conditions. Cotton fabrics inoculated by the microscopic fungus *Aspergillus niger* under laboratory conditions at a temperature +29°C lose 66% of the initial strength 2–3 weeks after contamination, whereas inoculation by *Chaetomium globosum* induces 98.7% loss of strength, that is completely destroys the material.

The same fabric exposed to soil at +29°C during 6 days loses 92% of the initial strength.

And cotton fabric exposed to sea water for 65 days loses up to 90% of strength.

1   2

FIGURE 10.32   *(Continued)*

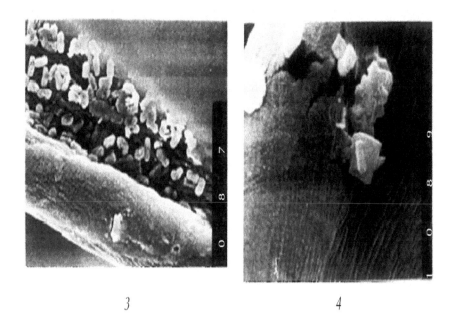

FIGURE 10.32   Cotton fiber micrographs: *1*—initial fiber (×4,500); *2–4:* fibers damages by different microorganisms: *2*—*Aspergillus niger* (×3,000); *3*—*Bacillus subtilis* (×4,500); *4*—*Pseudomonas fluorescens* (×10,000).

BAST FIBER BIODAMADING

Fibers produced from stalks, leaves or fruit covers of plants are called bast fibers. Hemp stalks give strong, coarse fibers—the hemp used for packing cloth and ropes. Coarse technical fibers: jute, ambary, ramie, and so on are produced from stalks of cognominal plants. Among all bast fibers, linen ones are most widely used.

The linen complex fiber, from which yarn and fabrics are manufactured, represents a batch of agglutinated filaments (plant cells) stretched and arrow-headed. The linen filament represents a plant cell with thick walls, narrow channel and knee-shaped nodes called shifts. Shifts are traces of fractures or bends of the fiber occurred during growth, and especially during mechanical treatment. Fiber ends are arrow-shaped, and the channel is closed. The cross-section represents an irregular polygon with five or six edges and a channel in the center. Coarser fibers have oval cross-section with wider and slightly flattened channel.

Complex fibers consist of filament batches (15–30 pieces in a batch) linked by middle lamellae. Middle lamellae consist of various substances: pectins, lignin, hemocellulase, and so on.

Bast fibers contain a bit lower amount of cellulose (about 70%) than cotton ones. Moreover, they contain such components as lignin (10%), wax and trace amounts of antibiotics, some of which increase biostability of the fiber. The presence of lignins induces coarsening (lignifications) of plant cells that promotes the loss of softness, flexibility, elasticity, and increased friability of fibers.

The main method for fiber separation from the flax is microbiological one, in which vital activity of pectin degrading microorganisms degrade pectins linking bast batches to the stalk tissues. After that the fiber can be easily detached by mechanical processing.

Microorganisms affect straw either at its spreading directly at the farm that lasts 20–30 days or at its retting at a flax-processing plant where retting lasts 2–4 days.

In the case of spreading and retting of spread straw by atmospheric fallouts and dew under anaerobic conditions, the main role is played by microscopic fungi. According to data by foreign investigators, the following fungi are the most widespread at straw spreading: *Pullularia* (spires in the stalk bark); *Cladosporium* (forms a velvet taint of olive to dark green color); *Alternaria* (grows through the bark by a flexible colorless chain and unambiguously plays an important role at dew spreading).

The studies indicate that *Cladosporium* fungus is the most active degrader of flax straw pectins.

When retting linen at flax-processing plants, conditions different from spreading are created for microflora. Here flax is submerged to the liquid with low oxygen content due to its displacement from straws by the liquid and consumption by aerobic bacteria, which propagate on easily accessible nutrients extracted from the straw.

These conditions are favorable for multiplication of anaerobic, pectin degrading clostridia related to the group of soil spore bacteria, which includes just few species. Most of them are thermophiles and, therefore, the process takes 2–4 days in the warmed up water; however, at lower temperature (+15…20°C) it takes 10–15 days.

CURIOUS FACTS

In Russia and Check Republic, spreading is the most popular way of processing flax. In Poland, Romania and Hungary, the flax is processed at flax-processing plants by retting, and I Netherlands—by retting and partly by spreading.

The linen fiber obtained by different methods (spread or retted straw) has different spinning properties. The spread straw is now considered to be the best, where the main role in degradation of stalk pectins is played by mold fungi. In production of retted fiber, this role is played by pectin degrading bacteria, some strains of which being able to form an enzyme (cellulase) that degrades cellulose itself. Such impact may be one of the damaging factors in the processes of linen retting. Thus, biostability of the flax depends on the method of fiber production.

The studies show that all kinds of biological treatment increase the quantity of various microbial damages of the fiber. Meanwhile, the spread fiber had lower total number of microscopic damages compared with any other industrial method.

There are other methods for flax production, steaming, for example, that gives steamed fiber. It has been found that steamed flax is the most biostable fiber. Possible reasons for so high biostability are high structure ordering of this fiber and high content of modified lignin in it. Moreover, during retting and spreading the fiber is enriched with microorganisms able to degrade cellulose under favorable conditions, whereas steaming sterilizes the fiber.

When exposed to microorganisms, pectins content in the linen fiber decreases by 38%, whereas cellulose content—by 1.2% only. The quantity of wax and ash content of the fiber exposed to microorganisms do not virtually change.

The ordered area share in the linen cellulose is 83.6%, the weakly ordered area—5.1%, and disordered area—15.7%. During microbiological degradation the share of disordered areas in the linen cellulose decreases to 7.8%, and the share of ordered areas increases to 86.9%. The share of weakly ordered areas varies insignificantly.

Microbiological damages of linen, jute and other bast fibers and fabrics are manifested by separate staining (occurrence of splotches of color or fiber darkening) and putrefactive odor. On damaged bast fibers, microscopic cross fractures and chips, and microholes and scabs in the fiber walls are observed.

The studies of relative biostability of bast fibers demonstrate that Manilla hemp and jute are most stable, whereas linen and cannabis fibers have the lowest stability.

Natural biostability of bast fibers is generally low and in high humidity and temperature conditions, when exposed to microorganisms, physicochemical and strength indices of both fibers and articles from them rapidly deteriorate. Generally, bast fibers are considered to have virtually the same biostability, as cotton fibers do.

Biostability of cellulose fivers is highly affected by further treatment with finishing solutions (sizing and finishing) containing starch, powder, resins and other substances which confer wearing capacity, wrinkle resistance, fire endurance, and so on to textile materials. Many of these substances represent a good culture medium for microorganisms. Therefore, at the stage of yarn and fabric sizing and finishing, the main attention is paid to strict compliance with sanitary and technological measures which are to prevent fabric infection by microorganisms and further biodamaging.

BIODAMAGING OF ARTIFICIAL FIBERS

Artificial fibers and fabrics are produced by chemical treatment of natural cellulose obtained from spruce, pine tree and fir. Artificial fibers based on cellulose are viscose, acetate, and so on ones. These fibers obtained from natural raw material have higher amorphous structure as compared with high-molecular natural material and, therefore, have lower stability, higher moisture and swelling capacity.

By chemical structure and microbiological stability viscose fibers are similar to common cotton fibers. Biostability of these fibers is low: many cellulosolytic microorganisms are capable of degrading them. Under laboratory conditions, some species of mold fungi shortly (within a month) induces complete degradation of viscose fibers, whereas wool fibers under the same conditions preserve up to 50% of initial stability. For viscose fabrics, the loss of stability induced by soil microorganisms during 12–14 days gives 54–76%. These parameters of artificial fibers and fabrics are somewhat higher than for cotton.

Acetate fibers are produced from acetyl cellulose—the product of cellulose etherification by acetic anhydride. Their properties significantly differ from those of viscose fibers and more resemble artificial fibers. For instance, they possess lower moisture retaining property, lesser swelling and loss of strength under wet condition. They are more stable to damaging effect of cellulosolytic enzymes of bacteria and microscopic fungi, because contrary to common cellulose fibers possessing side hydroxyl groups in macromolecules, acetate fiber macromolecules have side acetate groups hindering interaction of macromolecules with enzymes.

Among artificial textile materials of the new generation, textile fibers from bamboo, primarily obtained by Japanese, are highlighted. Bamboo possesses the reference antimicrobial properties due to the presence of "bambocane" substance in the fiber. Bamboo fivers possess extremely porous structure that makes them much more hygroscopic that cotton. Clothes from bamboo fibers struggle against sweat secretion—moisture is immediately absorbed and evaporated by fabric due to presence of pores, and high antimicrobial properties of bamboo prevent perspiration odor.

Various modified viscose fibers, micromodal and modal, for example, produced from beech were not studies for biostability. Information on biostability of artificial fibers produced from lactic casein, soybean protein, maize, peanut and corn is absent.

WOOL FIBER BIODAMAGING

By wool the animal hair is called, widely used in textile and light industry. The structure and chemical composition of the wool fiber significantly differ from other types of fibers and shows great variety and heterogeneity of properties. Sheep, camel, goat and rabbit wool is used as the raw material.

After thorough cleaning, the wool fiber can be considered virtually consisting of a single protein—keratin. The wool contains the following elements (in %): carbon—50; hydrogen—6–7; nitrogen—15–21; oxygen 21–24; sulfur—2–5, and other elements.

The chemical feature of wool is high content of various amino acids. It is known that wool is a copolymer of, at least, 17 amino acids, whereas the most of synthetic fibers represent copolymers of two monomers.

Different content of amino acids in wool fibers promotes the features of their chemical properties. Of the great importance is the quantity of cystine containing virtually all sulfur, which is extremely important for the wool fiber properties. The higher sulfur content in the wool is, the better its processing properties are, the higher resistance to chemical and other impacts is and the higher physic-mechanical properties are.

Wool fiber layers, in turn, differ by the sulfur content: it is higher in the cortical layer that in the core.

Among all textile fibers, wool has the most complex structure.

The fine merino wool fiber consists of two layers: external flaky layer or cuticle and internal cortical layer—the cortex. Coarser fibers have the third layer—the core.

The cuticle consists of flattened cells overlapping one another (the flakes) and tightly linked to one another and the cortical layer inside.

Cuticular cells have a membrane, the so-called epicuticle, right around. It is found that epicuticle gives about 2% of the fiber mass. Cuticle cells limited by walls quite tightly adjoin one another, but, nevertheless, there is a thin layer of intercellular protein substance between them, which mass is 3–4% of the fiber mass.

The cortical layer, the cortex, is located under the cuticle and forms the main mass of the fiber and, consequently, defines basic physico–mechanical and many other properties of the wool. Cortex is composed of spindle-shaped cells connivent to one another. Protein substance is also located between the cells.

The cortical layer cells are composed of densely located cylindrical, thread-like macrofibrils of about 0.05–0.2μm in diameter. Macrofibrils of the cortical layer are composed of microfibrils with the average diameter of 7–7.5nm.

Microfibrils, sometimes call the secondary agents, are composed of primary aggregates—protofibrils. Protofibril represents two or three twisted α-spiral chains.

It is suggested [27, 28] that α-spirals are twisted due to periodic repetition of amino acid residues in the chain, hence, side radicals of the same spiral are disposed in the inner space of another α-spiral providing strong interaction, including for the account of hydrophobic bonds, because each seventh residue has a hydrophobic radical.

According to the data by English investigator J. D. Leeder, the wool fiber can be considered as a collection of flaky and cortical cells bound by a cell membrane complex (CMC) which thus forms a uniform continuous phase in the keratine substance of the fiber. This intercellular cement can easily be chemically and microbiologically degraded, that is a δ-layer about 15nm thick (CMC or intercellular cement) is located between cells filling in all gaps.

The studies show that the composition of intercellular material between flaky cells may differ from that of the material between cortical cells. In the cuticle-cuticle, cuticle-cortex and cortex-cortex complexes the intercellular "cement" has different chemical compositions.

Although the cell membrane complex gives only 6% of the wool fiber mass, there are proofs that it causes the main effect on many properties of the fiber and fabric. For instance, a suggestion was made that CMC components may affect such mechanical properties, as wear resistance and torsion fatigue, as well as such chemical properties, as resistance to acids, proteolytic enzymes and chemical finishing agents.

The core layer is present in the fibers of coarser wool with the core cell content up to 15%. Disposition and shape of the core layer cells significantly vary with respect to the fiber type. This layer can be continuous (along the whole fiber) or may be separated in sections. The cell carcass of the core layer is composed of protein similar to microfibril cortex protein.

By its chemical composition, wool is a protein substance. The main substance forming wool is keratine—a complex protein containing much sulfur in contrast with other proteins. Keratine is produced during amino acid biosynthesis in the hair bag epidermis in the hide. Keratine structure represents a complex of high-molecular chain batches interacting both laterally and transversally. Along with keratine, wool contains lower amounts of other substances.

Wool keratine reactivity is defined by its primary, secondary, and tertiary structures, that is, the structure of the main polypeptide chains, the nature of side radicals and the presence of cross bonds.

Among all amino acids, only cystine forms cross bonds; their presence considerably defines wool insolubility in many reagents. Cystine bond decomposition simplifies wool damaging by sunlight, oxidants and other agents. Cystine contains almost all sulfur present in the wool fibers. Sulfur is very important for the wool quality, because it improves chemical properties, strength and elasticity of fibers.

Along with general regularities in the structure of high-molecular compounds, fibers differ from one another by chemical composition, monomer structure, polymerization degree, orientation, intermolecular bond strength and type, and so on that defines different physico–mechanical and chemical properties of the fibers.

The main chemical component of wool—keratine, is a nutrition for microorganisms.

Microorganisms may not directly consume proteins. Therefore, they are only consumed by microbes having proteolytic enzymes—exoproteases that are excreted by cells to the environment.

Wool damage may start already before sheepshearing, that is, in the fleece, where favorable nutritive (sebaceous matters, wax, and epithelium), temperature, aeration and humidity conditions are formed.

Contrary to microorganisms damaging plant fibers, the wool microflora is versatile, generally represented by species typical of the soil and degrading plant residues.

Initiated in the fleece, wool fiber damages are intensified during its storage, processing and transportation under unfavorable conditions.

Specific epiphytic microflora typical of this particular fiber is always present on its surface. Representatives of this microflora excrete proteolyric enzymes (mostly pepsin), which induce hydrolytic keratine decay by polypeptide bonds to separate amino acids.

Wool is degraded in several stages: first, microorganisms destroy the flaky layer and then penetrate into the cortical layer of the fiber, although the cortical layer itself is not destroyed, because intercellular substance located between the cells is the culture medium. As a result, the fiber structure is disturbed: flakes and cells are not bound yet, the fiber cracks and decays.

The mechanism of wool fiber hydrolysis by microorganisms suggested by American scientist E. Race represents a sequence of transformations: proteins—peptones—polypeptides—water + ammonia + carboxylic acids.

The most active bacteria: *Alkaligenes bookeri, Pseudomonas aeroginosa, Proteus vulgaris, Bacillus agri, B. mycoides, B. mesentericus, B. megatherium, B. subtilis,* and *microscopic fungi: Aspergillus, Alternaria, Cephalothecium, Dematium, Fusarium, Oospora, Penicillium, Trichoderma*, were extracted from the wool fiber surface.

However, the dominant role in the wool degradation is played by bacteria. Fungi are less active in degrading wool. Consuming fat and dermal excretion, fungi create conditions for further vital activity of bacteria-degraders. The role of microscopic fungi may also be reduced to splitting the ends of fibers resulting mechanical efforts of growing hyphae. Such splitting allows bacteria to penetrate into the fiber. Fungi weakly use wool as the source of carbon.

In 1960s, the data on the effect of fat and dirt present on the surface of unclean fibers on the wool biodamaging were published. It is found that unclean wool is damaged much faster than clean one. The presence of fats on unclean wool promotes fungal microflora development.

The activity of microbiological processes developing on the wool depends on mechanical damages of the fiber and preliminary processing of the wool.

It is found that microorganism penetration may happen through fiber cuts or microcracks in the flaky layer. Cracks may be of different origins—mechanical, chemical, and so on. It is also found that wool subject to intensive mechanical or chemical treatment is easier degraded by microorganisms than untreated one.

For instance, high activity of microorganisms during wool bleaching by hydrogen peroxide in the presence of alkaline agents and on wool washed in the alkaline medium was observed. When wool is treated in a weak acid medium, the activity of microorganisms is abruptly suppressed. This also takes place on the wool colored by chrome and metal-containing dyes. The middle activity of microorganisms is observed on the wool colored by acid dyes.

When impacted by microorganisms, structural changes in the wool are observed: flaky layer damages, its complete exfoliation, and lamination of the cortical layer.

Wool fiber damages can be reduced to several generalized types provided by their structural features:

- Channeling and overgrowth—accumulation of bacteria or fungal hyphae and their metabolites on the fiber surface;
- Flaky layer damage, local and spread;
- Cortical layer lamination to spindle-shaped cells;
- Spindle-shaped cell destruction.

Along with the fiber structure damage, some bacteria and fungi decrease its quality by making wool dirty blue or green that may not be removed by water or detergents. Splotches of color also occur on wool, for example, due to the impact of Pseudomonas aeruginosa bacteria; in this case, color depends on medium pH: green splotches occurred in a weakly alkaline medium, and in weak acid medium they are red. Green splotches may also be caused by development of Dermatophilus congolensis fungi. Black color of wool is provided by Pyronellaea glomerata fungi.

Thus, wool damage reduces its strength, increases waste quantity at combing and imparts undesirable blue, green or dirty color and putrefactive odor. However, wool is degraded by microorganisms slower than plant fibers.

CHANGES OF STRUCTURE AND PROPERTIES OF WOOL FIBERS BY MICROORGANISMS

To evaluate bacterial contamination of wool fibers, it is suggested to use an index suggested by A.I. Sapozhnikova, which characterizes discoloration rate of resazurin solution, a weak organic dye and currently hydrogen acceptor. It is also indicator of both presence and activity of reductase enzyme.

The method is based on resazurin ability to lose color in the presence of reductase, which is microorganisms' metabolite, due to redox reaction proceeding. This enables judging about quantity of active microorganisms present in the studied objects by solution discoloration degree.

Discoloration of the dye solution was evaluated both visually and spectrophotometrically by optical density value.

Table 10.1 shows results of visual observations of color transitions and optical density measurements of incubation solutions after posing the reductase test.

As follows from the data obtained, coloration of water extracts smoothly changed from blue-purple for control sterile physiological solution (D = 0.889) to purple for initial wool samples (D_{thin} = 0.821 and D_{coarse} = 0.779), crimson (D_{thin} = 0.657 and D_{coarse} = 0.651) and light crimson at high bacterial contamination (D_{thin} = 0.548 and D_{coarse} = 0.449 and 0.328) depending on bacterial content of the fibers.

TABLE 10.1 Visual coloration and optical density of incubation solutions with wool fibers at the wavelength λ = 600nm and different stages of spontaneous microflora development

Time of microorganism development, days	Thin merino wool		Coarse caracul wool	
	Optical density	Color (visual assessment)	Optical density	Color (visual assessment)
Control, physiological saline	0.889	Blue purple	0.889	Blue purple
0 (init.)	0.821	Purple	0.779	Purple
7	0.712	Purple	0.657	Crimson
14	0.651	Crimson	0.449	Light crimson
28	0.548	Light crimson	0.328	Light crimson

This dependence can be used for evaluation of bacterial contamination degree for wool samples applying color standard scale.

It is found that the impact of microorganisms usually reduces fiber strength, especially for coarse caracul wool: after 28 days of impact strength decreased by 57–65%. The average rate of strength reduction is about 2% per day (Figure 10.33).

It is found that after 28 days of exposure, the highest reduction of breaking load of wool fibers is induced by *Bac. subtilis* bacteria.

Figure 10.34 clearly shows that 14 days after exposure to microorganisms the surface of coarse wool fiber is almost completely covered by bacterial cells. Meanwhile, note also (Figure 10.35) that cuticular cells themselves are not damaged, but their bonding is disturbed that grants access to cortical cells for microorganisms. Figure 10.36 shows the wool fiber decay to separate fibrils caused by microorganisms.

Obr0 Isx Merinos 9/14/00 SEI   ├──── 30 μm ────┤

FIGURE 10.33 (*Continued*)

FIGURE 10.33 Micrographs of initial wool fibers (×1,000): a—thin merino wool; b—coarse caracul wool.

FIGURE 10.34 (*Continued*)

FIGURE 10.34 Micrographs of thin merino wool (a) and coarse caracul wool (b) after 14 days of exposure to *Bac. subtilis* (×1,000).

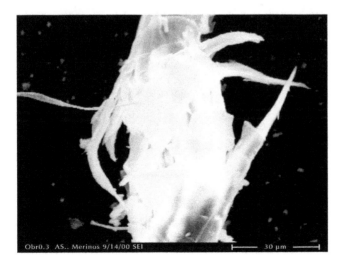

FIGURE 10.35 *(Continued)*

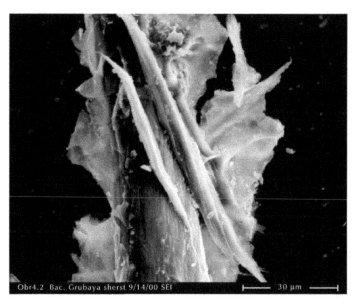

FIGURE 10.35 Micrographs of flaky layer destruction of thin (a) and coarse (b) wool fibers after 14 days of exposure to spontaneous microflora (×1,000).

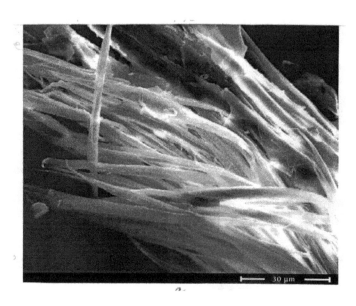

FIGURE 10.36 (*Continued*)

Obr4.1 Bac. Grubaya sherst 9/14/00 SEI   ├─── 30 μm ───┤ a

FIGURE 10.36   Micrographs of thin (a) and coarse (b) wool fiber fibrillation after 28 days of exposure to microorganisms (×1,000).

Wool fiber biodegradation changes the important quality indices, such as whiteness and yellowness. This process can be characterized as "yellowing" of wool fibers.

Table 10.4 shows data obtained on yellowness of thin merino and coarse caracul wool fibers exposed to spontaneous microflora, *Bac. subtilis* bacteria and *Asp. niger* fungi during 7, 14 and 28 days.

TABLE 10.4   Yellowness of thin merino and coarse caracul wool fibers after different times of exposure to different microorganisms

Time of exposure to microorganisms, days Type of fibers	Yellowness, % Impacting microorganisms		
	Bac. subtilis	Asp. niger	Spontaneous microflora
Thin, 0	27.7	27.7	27.7

TABLE 10.4   (*Continued*)

Thin, 7	36.4	37.3	29.1
Thin, 14	45.4	42.5	34.8
Thin, 28	53.9	48.1	39.9
Coarse 0	39.3	39.3	39.3
Coarse 7	46.8	44.2	41.2
Coarse 14	51.8	46.4	42.1
Coarse 28	57.1	49.5	43.3

In terms of detection of biodegradation mechanism and changes of material properties, determination of relations between changes in properties and structure of fibers affected by microorganisms is of the highest importance. Yellowness increase of wool fibers affected by microorganisms testifies occurrence of additional coloring centers.

For the purpose of elucidating the mechanism of microorganisms' action on the wool fibers that leads to significant changes of properties and structure of the material, the changes of amino acid composition of wool keratine proteins being the nutrition source of microorganisms damaging the material are studied.

Exposure to microorganisms during 28 days leads to wool fiber degradation and, consequently, to noticeable mass reduction of all amino acids in the fiber composition. To the greatest extent, these changes are observed for coarse wool, where total quantity of amino acids is reduced by 10–12rel.%, and for merino wool slightly lower reduction (4.7rel.%) is observed.

It should be noted that at comparatively low reduction of the average amount of amino acids in the system (not more than 12 rel.%) all types of wool fibers demonstrate a significant reduction of the quantity of some amino acids, such as serine, cystine, methionine, and so on (up to 25–33rel.%).

Analysis of the data obtained testifies that in all types of wool fibers (but to different extent) reduction of quantity of amino acids with disulfide bonds (cystine, methionine) and ones related to polar (hydrophilic) amino acids, including serine, glycine, threonine and tyrosine, is observed. These very amino acids provide hydrogen bonds imparting stability to keratine structure.

In the primary structure of keratine, serine is the N-end group, and Tyrosine is C-end group. In this connection, reduction of the quantity of these amino acids testifies degradation of the primary structure of the protein.

Changes observed in the amino acid composition of wool fiber proteins exposed to spontaneous microflora may testify that microorganisms degrade peptide and disulfide bonds, which provide stability of the primary structure of proteins, and break hydrogen bonds, which play the main role in stabilization of spatial structure of proteins (secondary, tertiary and quaternary).

Very important data were obtained in the study of the wool fiber structure by IR-spectroscopy method. It is found that when microorganisms affect wool fibers, their surface layers demonstrate increasing quantity of hydroxyl groups that indicates accumulation of functional COO-groups, nitrogen concentration in keratine molecule decreases, and protein chain configuration partly changes—β-configuration (stretched chains) transits to α-configuration (a spiral). This transition depends on α- and β-forms ratio in the initial fiber and is more significant for thin fibers, which mostly have β-configuration of chains in the initial fibers.

Thus, it is found that microorganisms generally affect the cell membrane complex and degrade amino acids, such as cystine, methionine, serine, glycine, threonine and tyrosine. Microorganisms reduce breaking load and causes "yellowing" of the wool fibers (Figure 10.37).

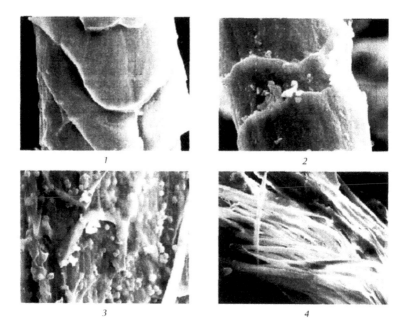

FIGURE 10.37   Micrographs of wool fibers: 1—initial undamaged fiber (×3,000); 2, 3—bacterial cells on the fiber surface (×3,000); 4—fiber fibrillation after exposure to microorganisms during 4 weeks (×1,000).

BIODAMAGES OF SYNTHETIC FIBERS

Synthetic fibers are principally different from natural and artificial ones by structure and, being an alien substrate for microorganisms, are harder damaged by them. Since occurrence of synthetic fabrics in 1950s, it is suggested that they are "everlasting" and are not utilized by microorganisms. However, it has been found with time that, firstly, microorganisms although slower, but yet are capable of colonizing synthetic fabrics and utilizing their carbon in the course of development (i.e. causing biodamage), and secondly, there are both more and less microorganism resistant fabrics among synthetic ones.

Among microorganisms damaging synthetic fibers, Trichoderma genus fungi are identified, at the initial stages developing due to lubricants and finishing agents without fiber damage and then wrap them with mycelium, loosen threads and, hence, reduce fabric strength.

When studying fabrics from nitrone, lavsan, caprone, it has been found that soil fungi and bacteria cause roughly the same effect on characteristics of these fabrics increasing the fiber swelling degree by 20–25%, reducing strength by 10–15% and elongation at break by 15–20%.

Synthetic fibers represent potential source of energy and nutrition for microorganisms. The ability of microorganisms to attach to surfaces of insoluble solids, then using them as the nutritive substrate, is well-known. Living cells of microorganisms have complex structure, just on the surface of bacterial cells complexes of proteins, lipids and polysaccharides were found; it contains hydrophilic and hydrophobic areas, various functional groups and mosaic electric charge (at total negative charge of the cells).

The first stage of microorganism interaction with synthetic fibers can be rightfully considered in terms of the adhesion theory with provision for the features of structure and properties of microorganisms as a biological system.

The entire process of microorganism impact of the fiber can conditionally be divided into several stages: attachment to the fiber, growth and multiplication on it and consumption of it, as the nutrition and energy source.

Enzymes excreted by bacteria act just in the vicinity of bacterial membrane. Been adsorbed onto the fiber, living cells attach to the surface and adapt to new living conditions. The ability to be adsorbed onto the surface of synthetic fibers is caused by:

- The features of chemical structure of the fibers. For instance, fibers adsorbing microorganisms are polyamide and polyvinyl alcohol ones; the fiber not adsorbing microorganisms is, for example, florin;
- Physical structure of the fiber. For example, fibers with smaller linear density, with a lubricant on the surface absorb greater amount of microorganisms;
- The presence of electric charge on the surface, its value and sign. Positively charged chemical fibers adsorb virtually all bacteria, fibers having no electric charge adsorb the majority of bacteria, and negatively charged fibers do not adsorb bacteria.

Supermolecular structure also stipulates the possibility for microorganisms and their metabolites to diffuse inside the internal areas of the fiber. Microorganism assimilation of the fiber starts from the surface, and further degradation processes and their rate are determined by microphysical state of the fiber. Microorganism metabolite penetration into inner areas of the fiber and deep layers of a crystalline material is only possible in the presence of capillaries.

Chemical fiber damaging and degradation starting from the surface are, in many instances, promoted by defects like cracks, chips or hollows which may occur in the course of fiber production and finishing.

Along with physical inhomogeneity, chemical inhomogeneity may promote biodegradation of synthetic fibers. Chemical inhomogeneity occurs during polymer synthesis and its thermal treatments, manifesting itself in different content of monomers and various end groups. The possibility for microorganism metabolites to penetrate inside the structure of synthetic fibers depends on the quantity and accessibility of functional end groups in the polymer, which are abundant in oligomers.

The ability to synthetic fibers to swell also makes penetration of biological agents inside low ordered areas of fibers and weakens intermolecular interactions, off-orientation of macromolecules, and degradation in the amorphous and crystalline zones. Structural changes result in reduction of strength properties of fibers.

Theoretical statements that synthetic fibers with the lower ordered structure and higher content of oligomers possess lower stability to microorganism impact than fibers with highly organized structure and lower content of low-molecular compounds.

Thus, the most rapid occurrence and biodegradation of synthetic fibers are promoted by low ordering and low orientation of macromolecules in the fibers, their low density, low crystallinity and the presence of defects in macro- and microstructure of the fibers, pores and cavities in their internal zones.

Carbochain polymer based fibers are higher resistant to microbiological damages. These polymers are: polyolefins, polyvinyl chloride, polyvinyl fluoride, polyacrylonitrile, polyvinyl alcohol. Fibers based on heterochain polymers: polyamide, polyether, polyurethane, and so on, are less bioresistant.

Comparative soil tests for biostability of artificial and synthetic fibers demonstrate that viscose fiber is completely destroyed on the 17th day of tests; bacterium and fungus colonies occur on lavsan on the 20th day; caprone is overgrown by fungus mycelium on the 30th day. Chlorin and

Torlon have the highest biostability. The initial signs of their biodamage are only observed 3 months after the test initiation.

The studies of nitron, lavsan and capron fabric biostability have found that soil fungi and bacteria cause nearly equal influence on parameters of these fabrics, increasing swelling degree of the fibers by 20–25%, reducing strength by 10–15% and elongation at break by 15–20%. Meanwhile, nitron demonstrated higher biostability, as compared with lavsan and capron.

CHANGES IN STRUCTURE AND PROPERTIES OF POLYAMIDE FIBERS INDUCED BY MICROORGANISMS

In contrast with natural fibers, chemical fibers have no permanent and particular microflora. Therefore, the most widespread species of microorganisms possessing increased adaptability are the main biodegraders of these materials.

Occurrence and progression of biological degradation of polyamide fibers is, in many instances, is induced by their properties and properties of affecting microorganisms, and their species composition. Generally, the species of microorganisms degrading polyamide and other chemical fibers are determined by their operation conditions, which form microflora, and its adaptive abilities.

Polyamide fibers are most frequently used in mixtures with natural fibers. Natural fibers contain specific microflora on the surface and inside. Therefore, capron fibers mixed with cotton, wool or linen are affected by their microflora. It is found that capron fiber degradation by microorganisms obtained from wool is characterized as deep fiber decay; microorganisms extracted from natural silk cause streakiness of capron fibers; microorganisms extracted from cotton cause fading and decomposition; microorganisms extracted from linen cause fading, streakiness and decomposition.

The microorganism interaction with polyamide fibers is most fully studied in the works by scientists. For the purpose of detecting bacteria-degraders of polyamide fibers, microorganisms were extracted from fibers damaged in the medium of active sewage silt, soil, microflora of natural fibers and test-bacteria complex selected as degraders of polyamide materials. Capron fibers were inoculated by extracted cultures of microorganisms and types of damages were reproduced.

Polyamide fiber materials were natural nutritive and energy source for these microorganisms. Therefore, bacterial strains extracted from damaged fibers were different from the initial strains. It is proven that the existence on a new substrate has stipulated changes of intensity and direction of physiological-biochemical processes of bacterial cells, the change of their morphological and culture properties.

Extraction, cultivation and use of such adaptive strains are of both scientific and practical interest. Using bacterial strains adaptive to capron, polyamide production waste, warn products, toxic substances may be utilized. This allows obtaining of secondary raw materials and solving the problem of environmental protection.

It is proved experimentally that polycaproamide fibers possess high adsorbability, which value depends on the properties of impacting bacteria. For instance, gram-positive, especially spore-forming bacteria *Bacillus subtilis, Bac. mesentericus* (from 84.5 to 99.3% of living cells), are most highly adsorbed, and adsorption of gram-positive bacteria varies significantly.

The extensive research of test-bacterium complex impact was performed. These bacteria were chosen as degraders of polyamide fibers, along with microflora of active sewage silts, linen and jute microorganisms as degraders of a complex capron thread. It is found that the test-bacterium complex injected by the author, after 7 months of exposure, increases biodegradation index to 1.27 that testifies about intensive degradation of the fiber microstructure. The highest degradation of complex polycaproamide (PCA) thread is caused by from active sewage silt microorganisms.

High activity of silt microorganisms and test-bacteria is explained by the fact that they include bacterium strains *Bacillus subtilis* and *Bac. mesentericus*, which according to the data by a number of authors may induce full degradation of caprolactam to amino acids using it, as the source of carbon and nitrogen.

Thus, adaptive forms of microorganisms induce the highest degradation of polyamide fibers.

Polyamide fibers are characterized by physical structure inhomogeneity, which occurs during processing and is associated with differences in crystallinity and orientation of macromolecules determining fiber accessibility for microorganisms and their metabolites penetration. The surface layer is damaged during orientational stretching and, consequently, has lower molecular alignment. That is why the surface layer is most intensively changed by microorganisms. The study of polyamide fiber macrostructure after exposure to microorganisms shows that streakiness and cover damage are the main damages of these fibers.

The studies of supermolecular structure of capron fiber surface show that after exposure to microorganisms the capron fiber cover becomes loose and uneven . Surface supermolecular structure degradation increases with the microbial impact: fibrils and their yarns become split and disaligned both laterally and transversely, multiple defects in the form of pores and cavities, and cracks of various depths are formed.

Chemical inhomogeneity of polyamide fibers also promotes changes in the fiber structure, when exposed to microorganisms. It is found that the polyamide fiber degradation increases with low-molecular compound (LMC) content in them; meanwhile, at the same content of low-molecular compounds, thermally treated fibers were higher biostable, as compared with untreated specimens.

Along with the morphological characters which characterize bio-damaging of the fibers, functional features, such as strength decrease and increase of deformation properties of the fiber, were detected. The greatest strength decrease (by 46.4%) was observed for thermally untreated capron fiber containing 3.4% LMC, and the smallest decrease (by 5%) was observed for the fibers with 3.2% LMC, thermally treated at the optimum time of 5s .

IR-spectroscopy method was applied to detect polycaproamide fiber damage by microorganisms [39]. It is found that carboxyl and amide groups are accumulated during their biodegradation.

The change of various property indices of polyamide fibers also results from macro-, micro- and chemical changes.

Of interest are studies of the microorganism effect on polyamide fabric quality. Fabrics (both bleached and colored) from capron monofilament were exposed to a set of test-cultures:

Bac.subtilis, Ps.fluorecsens, Ps.herbicola, Bac.mesentericus. After 3–9 month exposure, yellowness and dark spots occurred, coloring intensity decreased, and an odor appeared. The optical microscopy studies indicated that all fibers exposed to microorganisms, had damages typical of synthetic fibers—overgrowing, streakiness, bubbles, wall damages. The increasing quantity of biodamages with time results in tensile strength reduction: by 6–8% for capron fibers after 9 months, at inconsiderable change of relative elongation.

It all goes to show that development of microorganisms on polyamide fibrous materials results in changes of fibers morphology, their molecular and supermolecular structure and, as a consequence, reduction of strength properties, color change, and odor.

To clear up the mechanism of polycaproamide fiber degradation by polarographic investigation, a possibility of ε-amino caproic acid accumulation by *Bacillus subtilis k1* culture during degradation of polycaproamide fibers 0.3 and 0.7tex fineness was studied (Figure 10.38).

As a source of carbon, ε-amino caproic acid (10mg/l) or PCA (0.5g/l) was injected into the mineral medium.

It is found that at PCA fibrous materials exposure to *Bacillus subtilis k1* strain, the maximum quantity of ε-amino caproic acid is liberated on the fifth day: 32mg/l, 66mg/l.

Amino caproic acid accumulation, mg\L:

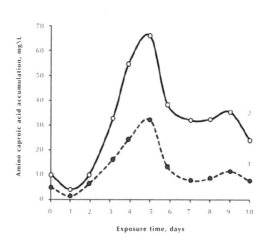

FIGURE 10.38   The change of ε-amino caproic acid (ACA) concentration during *Bacillus subtilis k1* development on PCA material fibers:-0.7tex; -0.3tex.

It is known that if the medium includes several substrates metabolized by a particular strain of microorganisms, the substrate providing the maximum culture propagation rate is consumed first. As this substrate is going to exhaust, bacteria subsequently consume other substrates, which provide lower rates of cell multiplication.

In the mineral medium containing chemically pure ε-amino caproic acid (with the initial concentration of 10mg/l) as the source of carbon, its concentration decreased gradually, and 5 days after it was not detected in the solution.

Thus, basing on the data obtained, one may conclude that, firstly, PCA fibers of 0.3tex fineness are more accessible for microorganisms; secondly, *Bacillus subtilis k1* strain can be used to utilize polycaproamide fibrous materials; thirdly, polarographic analysis has proven the mechanism of PCA fibrous material degradation with ε-amino caproic acid liberation. As a consequence, suggested *Bacillus subtilis k1* strain VKM No.V-1676D degrades PCA fibrous materials at the both macro- and microstructure levels, with ε-amino caproic acid formation.

METHODS OF TEXTILE MATERIAL PROTECTION AGAINST DAMAGING BY MICROORGANISMS

Imparting antimicrobial properties to textile materials pursues two main aims: protection of the objects contacting with textile materials against actions of microorganisms and pathogenic microflora.

In the first case, we speak about imparting biostability to materials and, as a consequence, about passive protection. The second case concerns creation of conditions for preventive attack of a textile material on pathogenic bacteria and fungi to prevent their impact on the protected object.

The basic method of increasing biostability of textile materials is application of antimicrobial agents (biocides). The requirements to the "ideal" biocide are the following:

- The efficacy against the most widespread microorganisms at minimal concentration and maximal action time;
- Non toxicity of applied concentrations for people;
- The absence of color and odor;

- Low price and ease of application;
- Retaining of physico–mechanical, hygienic and other properties of the product;
- Compatibility with other finishing agents and textile auxiliaries;
- light stability and weather ability.

At any time, nearly every class of chemical compounds was applied to impart textile materials antibacterial or antifungal activity. Today, application of nanotechnologies, specifically injection of silver and iodine nanoparticles, to impart textile antimicrobial properties is of the greatest prospect.

At all times, copper, silver, tin, mercury, and so on salts were used to protect fibrous materials against biodamages. Among these biocides, the most widespread are copper salts due to their low cost and comparatively toxicity. The use of zinc salts is limited by their low biocide action, whereas mercury, tin and arsenic salts are highly toxic for the man. However, there are organomercury preparations applied to synthetic and natural fibrous materials used as linings and shoe plates, widely advertised for antibacterial and antifungal finishing.

The data are cited that impregnation of textile materials by a mixture of neomycin with tartaric, propionic, stearic, phthalic and some other acids imparts them the bacteriostatic effect. Acids were dissolved in water, methyl alcohol or butyl alcohol and were sprayed on the material.

The methods of imparting textile materials biostability can be divided into the following groups:
- Impregnation by biocides, chemical and physical modification of fibers and threads, which then form a textile material;
- Cloth impregnation by antimicrobial agent solutions of emulsions, its chemical modification;
- Injection of antimicrobial agents into the binder (at nonwoven material manufacture by the chemical method);
- Imparting antimicrobial properties to textile materials during their coloring and finishing;
- Application of disinfectants during chemical cleaning or laundry of textile products.

However, impregnation of fibers and cloth does not provide firm attachment of reagents. As a result, the antimicrobial action of such materials is nondurable. The most effective methods of imparting biocide properties to textile materials are those providing chemical bond formation, that is chemical modification methods. Chemical modification methods for fibrous materials represent processing that leads to clathrate formation, for example injection of biological active agents into spinning melts or solutions.

At the stage of capron polymerization, an antibacterial organotin compound (tributyltin oxide or hydroxide) is added that retains the antibacterial effect after multiple laundries. Methods of imparting antimicrobial properties to textile materials by injection of nitrofuran compounds into spinning melts with further fixing them at molding in the fine structure of fibers similar to clathrates were designed.

There are data on imparting antimicrobial properties to synthetic materials during oiling. Prior to drafting, fibers are treated by compounds based on oxyquinoline derivatives, by aromatic amines or nitrofuran derivatives. Such fibers possess durable antimicrobial effect.

Nanotechnologies are actively intruded in the light industry, allowing obtaining of materials with antimicrobial properties. The following directions using nanotechnologies, which are now investigated, should be outlined, including creation of new textile materials on the account of:

• Primarily, the use of textile nonfibers and threads in the materials;
• Secondly, the use of nanodispersions and nanoemulsions for textile finishing.

It may be said that nanotechnologies allow a significant decrease of expenses at the main production stage, where consumption of raw materials and semiproducts is considerable. For nanoparticles imparting antimicrobial properties, silver, copper, palladium, and so on particles are widely used. Silver is the natural antimicrobial agent, which properties are intensified by the nanoscale of particles (the surface area sharply increases), so that such textile is able to kill multiple microorganisms and viruses. In this form, silver also reduces necessity in fabric cleaning, eliminates sweat odor as a result of microorganism development on the human body during wearing.

Properties of materials designed with the use of silver nanoparticles, which prevent multiplication of various microorganisms may be useful in medicine, for example. The examples are surgical retention sutures, bandages, plasters, surgical boots, medical masks, whites, skull-caps, towels, and so on. Customers know well sports clothes, prophylactic socks with antimicrobial properties.

Along with chemical fibers and threads, natural ones are treated by nanoparticles. For example, silver and palladium nanoparticles (5–20nm in diameter) were synthesized in citric acid, which prevented their agglutination and then natural fibers were dipped in the solution with these negatively charged particles. Nanoparticles imparted antibacterial properties and even ability to purify air from pollutants and allergens to clothes and underwear.

When these products appeared at the world market, disputes about ecological properties and the influence of these technologies on the human organism have arisen. There are no accurate data yet how these developments may affect the human organism. However, it should be noted that some specialists do not recommend everyday use of antibacterial socks, because these antibacterial properties affect the natural skin microflora.

Nanomaterials are primarily hazardous due to their microscopic size. Firstly, owing to small size they are chemically more active because of a great total area of the nanosubstance. As a result, low toxic substance may become extremely toxic. Secondly, chemical properties of the nanosubstance may significantly change due to manifestations of quantum effects that, finally, may make a safe substance extremely hazardous. Thirdly, due to small size, nanoparticles freely permeate through cellular membranes damaging bioplasts and disturbing the cell operation.

Physical modification of fibers or threads is the direct change of their composition (without new chemical formations and transformations), structure (supermolecular and textile), properties, production technology and processing. Modernization of the structure and increase of the fiber crystallinity degree induces biostability increase. However, in contrast with chemical modification, physical modification does not impart antimicrobial properties to the fibers, but may increase biostability.

By no means always textile materials produced completely from antimicrobial fibers are required. Even a small fracture of highly active antimicrobial fiber (e.g. 1/3 or even 1/4) is able to provide sufficient biostability to the entire material. The studies show that antimicrobial fibers were found not only protected themselves against microorganism damage, but also capable of shielding plant fibers from their impact.

Manufacture of antimicrobial nonwoven materials by injection of active microcapsule ingredients into it is of interest. Microcapsules can contain solid particles of microdrops of antimicrobial substances liberated under particular conditions (e.g. by friction, pressure, dissolution of capsule coatings or their biodegradation).

Biostability of fibrous materials may be significantly affected by the dye selection. Dyes possessing antimicrobial activity on the fiber are known: salicylic acid derivatives capable of bonding copper, triphenylmethane, acridic, thiazonic, and so on ones. For instance, chromium-containing dyes possess antibacterial action, but resistance to mold fungi is not imparted.

It is known that synthetic fibers dyed by dispersed pigments are more intensively degraded by microorganisms. It is suggested that these pigments make the fiber surface more accessible for bacteria and fungi.

Single bath coloring and bioprotective finishing of textile materials are also applied. A combination of these processes is not only of theoretical interest, but is also perspective in terms of technology and economy.

Processing of textile materials by silicones also imparts antimicrobial properties to these clothes. Some authors state that textile material sizing by water repellents imparts them sufficient antimicrobial activity. Water repellency of materials may reduce the adverse impact of microorganisms, because the quantity of adsorbed moisture is reduced. However, Hydrophobic finishing itself may not fully eliminate the adverse effect of microorganisms. Therefore, antimicrobial properties imparted to some textile materials during silicone finishing may be related to application of metal salts as catalysts, such as copper, chromium and aluminum.

Disinfectants, for example at laundry, may be applied by the customer himself. The method of sanitizing substance application for carpets, which is spraying or dispensing of a disinfectant on the surface of floor covers during operation is known. The acceptable disinfection level may be obtained during laundry of textile products by such detergents, which may create residual fungal and bacteriostatic activity.

APPENDIX B

Figure 10.39 shows the scanning electronic micrographs of carbon chitosan/nanotube electrospun fibers, at different chitosan concentrations, in a TFA/DCM (70:30) solvent. As presented in Figure 10.39a, at low concentrations of chitosan, the beads deposited on the collector and the thin fibers coexisted. As the concentration of chitosan increased (Figure 10.39a–c), the beads decreased significantly. Figure 10.39c shows homogenous electrospun fibers with minimum beads, thin fibers and interconnected fibers. The increase of chitosan concentration leads to an increase of the interconnected fibers, as shown in Figure 10.39d–e. The average diameter of chitosan/carbon nanotube fibers increased when increasing the concentration of chitosan (Figure 10.39a–e). Hence, a chitosan/carbon nanotube solution of TFA/DCM (70:30) with 10 wt.% of chitosan assured optimized conditions for electrospinning of this solution. When the voltage was low, the beads were deposited on the collector (Figure 10.40a). As shown in Figure 10.40a–d, the number of beads decreased with increasing voltage from 18 to 24kV. In our study, the average diameter of fibers prepared by 18kV was measured as 307nm. As the applied voltage increased, the average fiber diameters also increased. The average diameter of fibers for 20kV (10.40b), 22kV(10.40c), and 24kV (10.40d), was 308 (194–792), 448 (267–656) and 455 (306–672), respectively.

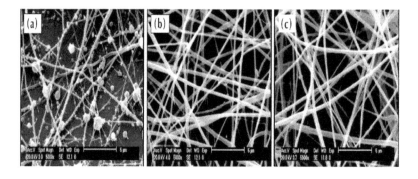

FIGURE 10.39   (*Continued*)

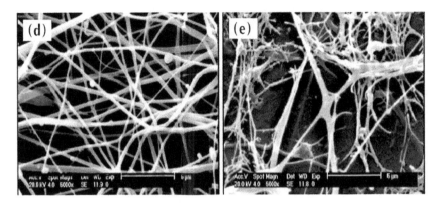

FIGURE 10.39 Scanning electron micrographs of electrospun fibers at different chitosan concentrations (wt.%): (a) 8, (b) 9, (c) 10, (d) 11, (e) 12, 24kV, 5cm, TFA/DCM: 70/30.

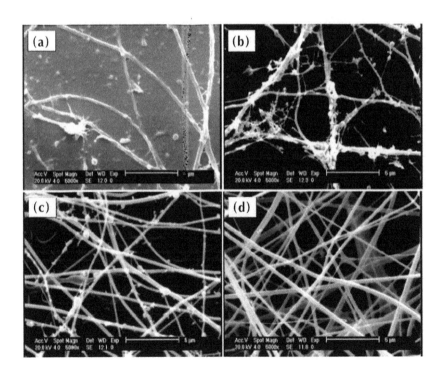

FIGURE 10.40 Scanning electron micrographs of electrospun fibers at different voltages (kV): (a) 18, (b) 20, (c) 22, (d) 24, 5cm, 10 wt.%, TFA/DCM: 70/30.

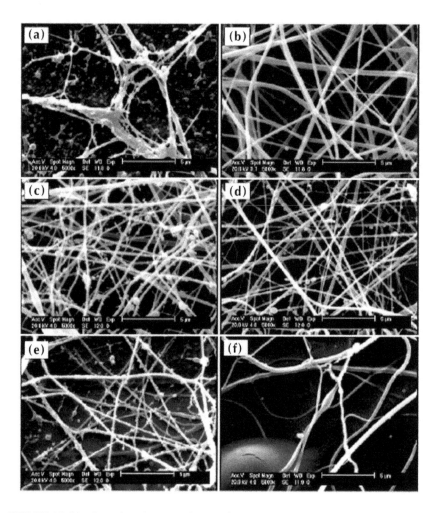

FIGURE 10.41 Scanning electron micrographs of electrospun fibers of chitosan/carbon nanotubes at different tip-to-collector distances (cm): (a) 4, (b) 5, (c) 6, (d) 7, (e) 8, (f) 10, 24kV, 10 wt.%, TFA/DCM: 70/30.

The morphologies of chitosan/carbon nanotube electrospun fibers at different tip-to-collector distances are presented in Figure 10.41. When the tip-to-collector distance was low, a little interconnected fiber (with high fiber diameter) was deposited on the collector (as shown in Figure 10.41a). At a 5cm tip-to-collector distance (Figure 10.41b), more homogenous fibers with negligible beads were obtained. However, the beads increased with increasing the tip-to-collector distance (Figure 10.41b–f).

Also, our study has demonstrated that the diameter of electrospun fibers decreased by increasing tip-to-collector distance (as shown in Figure 10.41b–d; 455 (306–672), 134 (87–163), 107 (71–196)). The fibers prepared within a distance of 8cm (Figure 10.41e) and 10cm (Figure 10.41f) presented defects and non homogenous diameter. However, a 5-cm tip-to-collector distance appears to be reliable for electrospinning.

APPENDIX C

In this appendix, effects of four electrospinning parameters, including solution concentration (wt.%), applied voltage (kV), tip to collector distance (cm), and volume flow rate (ml/hr), on contact angle (CA) of polyacrylonitrile (PAN) nanofiber mat are presented. To optimize and predict the CA of electrospun fiber mat, response surface methodology (RSM) and artificial neural network (ANN) are employed and a quantitative relationship between processing variables and CA of electrospun fibers is established.

MEASUREMENT AND CHARACTERIZATION

The morphology of the gold-sputtered electrospun fibers were observed by scanning electron microscope (SEM, Philips XL-30). The average fiber diameter and distribution was determined from selected SEM image by measuring at least 50 random fibers. The wettability of electrospun fiber mat was determined by CA measurement. The CA measurements were carried out using specially arranged microscope equipped with camera and PCTV vision software as shown in Figure 10.2. The droplet used was distilled water and was $1\mu l$ in volume. The CA experiments were carried out at room temperature and were repeated five times. All contact angles measured within 20s of placement of the water droplet on the electrospun fiber mat. A typical SEM image of electrospun fiber mat, its corresponding diameter distribution and CA image are shown in Figures 10.42 and 10.43.

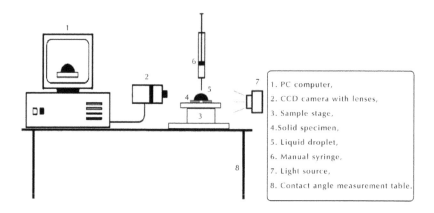

FIGURE 10.42 Schematic of CA measurement set up.

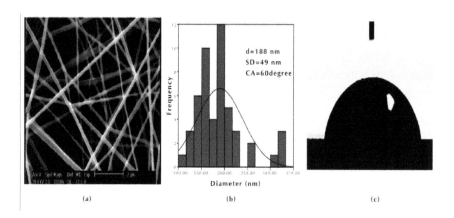

FIGURE 10.43 A typical (a) SEM image, (b) fiber diameter distribution, and (c) CA of electrospun fiber mat.

RESPONSE SURFACE METHODOLOGY

Response surface methodology (RSM) is a combination of mathematical and statistical techniques used to evaluate the relationship between a set of controllable experimental factors and observed results. This optimization process is used in situations where several input variables influence some output variables (responses) of the system.

In the present study, central composite design (CCD) was employed to establish relationships between four electrospinning parameters and the CA of electrospun fiber mat. The experiment was performed for at least three levels of each factor to fit a quadratic model. Based on preliminary experiments, polymer solution concentration (wt.%), applied voltage (kV), tip to collector distance (cm), and volume flow rate (ml/hr) were determined as critical factors with significance effect on CA of electrospun fiber mat. These factors were four independent variables and chosen equally spaced, while the CA of electrospun fiber mat was dependent variable. The values of -1, 0, and 1 are coded variables corresponding to low, intermediate and high levels of each factor respectively. The experimental parameters and their levels for four independent variables are shown in Table 10.5. The regression analysis of the experimental data was carried out to obtain an empirical model between processing variables. The contour surface plots were obtained using Design-Expert software.

TABLE 10.5 Design of experiment (factors and levels)

Factor	Variable	Unit	Factor level		
			-1	0	1
X1	Solution concentration	(wt.%)	10	12	14
X2	Applied voltage	(kV)	14	18	22
X3	Tip to collector distance	(cm)	10	15	20
X4	Volume flow rate	(ml/hr)	2	2.5	3

The quadratic model, Equation (10.2) including the linear terms, was fitted to the data.

$$Y = \beta_0 + \sum_{i=1}^{4} \beta_i x_i + \sum_{i=1}^{4} \beta_{ii} x_i^2 + \sum_{i=1}^{3} \sum_{j=2}^{4} \beta_{ij} x_i x_j \qquad (10.3)$$

where, x_i is the predicted response, x_i and x_j are the independent variables, β_0 is a constant, β_i is the linear coefficient, β_{ii} is the squared coefficient, and β_{ij} is the second-order interaction coefficients.

The quality of the fitted polynomial model was expressed by the determination coefficient (R^2) and its statistical significance was performed with the Fisher's statistical test for analysis of variance (ANOVA).

ARTIFICIAL NEURAL NETWORK

Artificial neural network (ANN) is an information processing technique, which is inspired by biological nervous system, composed of simple unit (neurons) operating in parallel. A typical ANN consists of three or more layers, comprising an input layer, one or more hidden layers and an output layer. Every neuron has connections with every neuron in both the previous and the following layer. The connections between neurons consist of weights and biases. The weights between the neurons play an important role during the training process. Each neuron in hidden layer and output layer has a transfer function to produce an estimate as target. The interconnection weights are adjusted, based on a comparison of the network output (predicted data) and the actual output (target), to minimize the error between the network output and the target.

In this study, feed forward ANN with one hidden layer composed of four neurons was selected. The ANN was trained using back-propagation algorithm. The same experimental data used for each RSM designs were also used as the input variables of the ANN. There are four neurons in the input layer corresponding to four electrospinning parameters and one neuron in the output layer corresponding to CA of electrospun fiber mat. Figure 10.44 illustrates the topology of ANN used in this investigation.

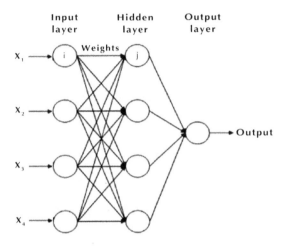

FIGURE 10.44   The topology of artificial neural network used in this study.

THE ANALYSIS OF VARIANCE (ANOVA)

All 30 experimental runs of CCD were performed according to Table 10.6. A significance level of 5% was selected; that is, statistical conclusions may be assessed with 95% confidence. In this significance level, the factor has significant effect on CA if the p-value is less than 0.05. On the other hand, when p-value is greater than 0.05, it is concluded the factor has no significant effect on CA.

The results of analysis of variance (ANOVA) for the CA of electrospun fibers are shown in Table 10.7. Equation (10.4) is the calculated regression equation.

TABLE 10.6   The actual design of experiments and response

No.	Electrospinning parameters				Response
	X_1	X_2	X_3	X_4	CA (°)
	Concentration	Voltage	Distance	Flow rate	
1	10	14	10	2	44 ± 6
2	10	22	10	2	54 ± 7

TABLE 10.6   (*Continued*)

3	10	14	20	2	61 ± 6
4	10	22	20	2	65 ± 4
5	10	14	10	3	38 ± 5
6	10	22	10	3	49 ± 4
7	10	14	20	3	51 ± 5
8	10	22	20	3	56 ± 5
9	10	18	15	2.5	48 ± 3
10	12	14	15	2.5	30 ± 3
11	12	22	15	2.5	35 ± 5
12	12	18	10	2.5	22 ± 3
13	12	18	20	2.5	30 ± 4
14	12	18	15	2	33 ± 4
15	12	18	15	3	25 ± 3
16	12	18	15	2.5	26 ± 4
17	12	18	15	2.5	29 ± 3
18	12	18	15	2.5	28 ± 5
19	12	18	15	2.5	25 ± 4
20	12	18	15	2.5	24 ± 3
21	12	18	15	2.5	21 ± 3
22	14	14	10	2	31 ± 4
23	14	22	10	2	35 ± 5
24	14	14	20	2	33 ± 6
25	14	22	20	2	37 ± 4
26	14	14	10	3	19 ± 3
27	14	22	10	3	28 ± 3
28	14	14	20	3	39 ± 5
29	14	22	20	3	36 ± 4
30	14	18	15	2.5	20 ± 3

$$CA = 25.80 - 9.89X_1 + 2.17X_2 + 4.33X_3 - 2.33X_4$$
$$- 1.63X_1X_2 - 1.63X_1X_3 + 1.63X_1X_4 - 0.88X_2X_3 - 0.63X_2X_4 + 0.37X_3X_4 \quad (10.4)$$
$$+ 7.90X_1^2 + 6.40X_2^2 - 0.096X_3^2 + 2.90X_4^2$$

TABLE 10.7 Analysis of variance for the CA of electrospun fiber mat

Source	SS	DF	MS	F-value	Probe > F	Remarks
Model	4175.07	14	298.22	32.70	<0.0001	Significant
X1-Concentration	1760.22	1	1760.22	193.01	<0.0001	Significant
X2-Voltage	84.50	1	84.50	9.27	0.0082	Significant
X3-Distance	338.00	1	338.00	37.06	<0.0001	Significant
X4-Flow rate	98.00	1	98.00	10.75	0.0051	Significant
X1X2	42.25	1	42.25	4.63	0.0481	Significant
X1X3	42.25	1	42.25	4.63	0.0481	Significant
X1X4	42.25	1	42.25	4.63	0.0481	Significant
X2X3	12.25	1	12.25	1.34	0.2646	
X2X4	6.25	1	6.25	0.69	0.4207	
X3X4	2.25	1	2.25	0.25	0.6266	
X_1^2	161.84	1	161.84	17.75	0.0008	Significant
X_2^2	106.24	1	106.24	11.65	0.0039	Significant
X_3^2	0.024	1	0.024	0.0026	0.9597	
X_4^2	21.84	1	21.84	2.40	0.1426	
Residual	136.80	15	9.12			
Lack of Fit	95.30	10	9.53	1.15	0.4668	

From the p-values presented in Table 10.7, it can be concluded that the p-values of terms terms 23X, 24X, 32XX,42XX and X_3X_4 is greater than the significance level of 0.05, therefore they have no significant effect on the CA of electrospun fiber mat. Since the above terms had no significant effect on CA of electrospun fiber mat, these terms were removed. The fitted equations in coded unit are given in Equation (10.5).

$$CA = 26.07 - 9.89X_1 + 2.17X_2 + 4.33X_3 - 2.33X_4$$
$$-1.63X_1X_2 - 1.63X_1X_3 + 1.63X_1X_4$$
$$+9.08X_1^2 + 7.58X_2^2$$

(10.5)

Now, all the p-values are less than the significance level of 0.05. Analysis of variance showed that the RSM model was significant (p<0.0001), which indicated that the model has a good agreement with experimental data. The determination coefficient (R^2) obtained from regression equation was 0.958.

ARTIFICIAL NEURAL NETWORK

In this section, the best prediction, based on minimum error, was obtained by ANN with one hidden layer. The suitable number of neurons in the hidden layer was determined by changing the number of neurons.

The good prediction and minimum error value were obtained with four neurons in the hidden layer. The weights and bias of ANN for CA of electrospun fiber mat are given in Table 10.8. The R^2 and mean absolute percentage error were 0.965 and 5.94% respectively, which indicates that the model was shows good fitting with experimental data.

TABLE 10.8   Weights and bias obtained in training ANN

Hidden layer	Weights	IW_{11}	IW_{12}	IW_{13}	IW_{14}
		1.0610	1.1064	21.4500	3.0700
		IW_{21}	IW_{22}	IW_{23}	IW_{24}
		-0.3346	2.0508	0.2210	-0.2224
		IW_{31}	IW_{32}	IW_{33}	IW_{34}
		-0.6369	-1.1086	-41.5559	0.0030
		IW_{41}	IW_{42}	IW_{43}	IW_{44}
		-0.5038	-0.0354	0.0521	0.9560
	Bias	b_{11}	b_{21}	b_{31}	b_{41}
		-2.5521	-2.0885	-0.0949	1.5478
Output layer	Weights	LW_{11}			
		0.5658			
		LW_{21}			
		0.2580			
		LW_{31}			
		-0.2759			
		LW_{41}			
		-0.6657			
	Bias	b			
		0.7104			

EFFECTS OF SIGNIFICANT PARAMETERS ON RESPONSE

The morphology and structure of electrospun fiber mat, such as the nanoscale fibers and interfibrillar distance, increases the surface roughness as well as the fraction of contact area of droplet with the air trapped between fibers. It is proved that the CA decrease with increasing the fiber diameter, therefore the thinner fibers, due to their high surface roughness, have higher CA than the thicker fibers. Hence, we used this fact for comparing CA of electrospun fiber mat. The interaction contour plot for CA of electrospun PAN fiber mat are shown in Figure 10.45.

As mentioned in the literature, a minimum solution concentration is required to obtain uniform fibers from electrospinning. Below this concentration, polymer chain entanglements are insufficient and a mixture of beads and fibers is obtained. On the other hand, the higher solution concentration would have more polymer chain entanglements and less chain mobility. This causes the hard jet extension and disruption during electrospinning process and producing thicker fibers. Figure 10.45(a) show the effect of solution concentration and applied voltage at middle level of distance (15cm) and flow rate (2.5ml/hr) on CA of electrospun fiber mat. It is obvious that at any given voltage, the CA of electrospun fiber mat decrease with increasing the solution concentration.

Figure 10.45(b) shows the response contour plot of interaction between solution concentration and spinning distance at fixed voltage (18kV) and flow rate (2.5ml/hr). Increasing the spinning distance causes the CA of electrospun fiber mat to increase. Because of the longer spinning distance could give more time for the solvent to evaporate, increasing the spinning distance will decrease the nanofiber diameter and increase the CA of electrospun fiber mat. As demonstrated in Figure 10.45(b), low solution concentration cause the increase in CA of electrospun fiber mat at large spinning distance.

The response contour plot in Figure 10.45(c) represented the CA of electrospun fiber mat at different solution concentration and volume flow rate. Ideally, the volume flow rate must be compatible with the amount of solution removed from the tip of the needle. At low volume flow rates, solvent would have sufficient time to evaporate and thinner fibers were produced, but at high volume flow rate, excess amount of solution fed to the tip of needle and thicker fibers were resulted. Therefore the CA of electrospun fiber mat will be decreased.

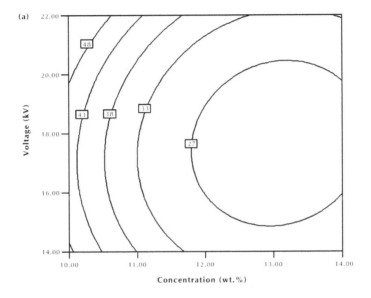

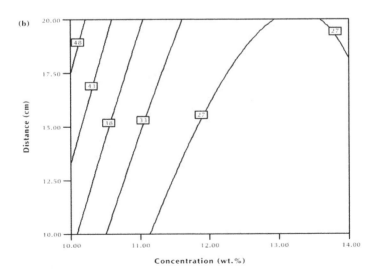

Figure 10.45   (*Continued*)

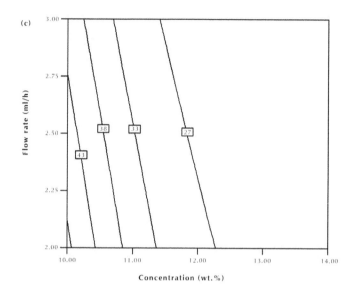

FIGURE 10.45 Contour plots for contact angle of electrospun fiber mat showing the effect of: (a) solution concentration and applied voltage, (b) solution concentration and spinning distance, (c) solution concentration and volume flow rate.

As shown by Equation (10.6), the relative importance (RI) of the various input variables on the output variable can be determined using ANN weight matrix.

$$RI_j = \frac{\sum_{m=1}^{N_h}\left(\left(\left|IW_{jm}\right|\middle/\sum_{k=1}^{N_i}\left|IW_{km}\right|\right)\times\left|LW_{mn}\right|\right)}{\sum_{k=1}^{N_i}\left\{\sum_{m=1}^{N_h}\left(\left(\left|IW_{km}\right|\middle/\sum_{k=1}^{N_i}\left|IW_{km}\right|\right)\times\left|LW_{mn}\right|\right)\right\}}\times 100 \qquad (10.6)$$

where RI_j is the relative importance of the jth input variable on the output variable, N_i and N_h are the number of input variables and neurons in hidden layer, respectively ($N_i = 4$, $N_h = 4$ in this study), IW and LW are the connection weights, and subscript "n" refer to output response ($n = 1$).

The relative importance of electrospinning parameters on the value of CA calculated by Equation (10.6) and is shown in Figure 10.46. It can be seen that, all of the input variables have considerable effects on the CA of electrospun fiber mat. Nevertheless, the solution concentration with relative importance of 49.69% is found to be most important factor affecting the CA of electrospun nanofibers. These results are in close agreement with those obtained with RSM.

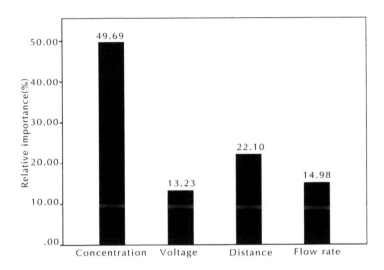

FIGURE 10.46 Relative importance of electrospinning parameters on the CA of electrospun fiber mat.

OPTIMIZING THE CA OF ELECTROSPUN FIBER MAT

The optimal values of the electrospinning parameters were established from the quadratic form of the RSM. Independent variables (solution concentration, applied voltage, spinning distance, and volume flow rate) were set in range and dependent variable (CA) was fixed at minimum. The optimal conditions in the tested range for minimum CA of electrospun fiber mat are shown in Table 10.9. This optimum condition was a predicted value, thus to confirm the predictive ability of the RSM model for response, a further electrospinning and CA measurement was carried out according to the optimized conditions and the agreement between predicted and measured responses was verified. Figure 10.47 shows the SEM, average fiber diameter distribution and corresponding CA image of electrospun fiber mat prepared at optimized conditions.

TABLE 10.9   Optimum values of the process parameters for minimum CA of electrospun fiber mat

Solution concentra- tion (wt.%)	Applied voltage (kV)	Spinning distance (cm)	Volume flow rate (ml/hr)	Pre- dicted CA (°)	Observed CA (°)
13.2	16.5	10.6	2.5	20	21

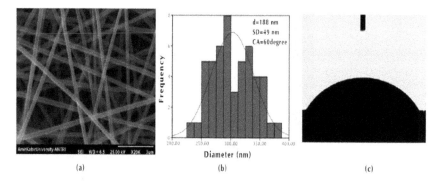

(a)   (b)   (c)

FIGURE 10.47   (a) SEM image, (b) fiber diameter distribution, and (c) CA of electrospun fiber mat prepared at optimized conditions.

COMPARISON BETWEEN RSM AND ANN MODEL

Table 10.10 gives the experimental and predicted values for the CA of electrospun fiber mat obtained from RSM as well as ANN model. It is demonstrated that both models performed well and a good determination coefficient was obtained for both RSM and ANN. However, the ANN model shows higher determination coefficient ($R^2 = 0.965$) than the RSM model ($R^2 = 0.958$). Moreover, the absolute percentage error in the ANN prediction of CA was found to be around 5.94%, while for the RSM model, it was around 7.83%. Therefore, it can be suggested that the ANN model shows more accurately result than the RSM model. The plot of actual and predicted CA of electrospun fiber mat for RSM and ANN is shown in Figure 10.48.

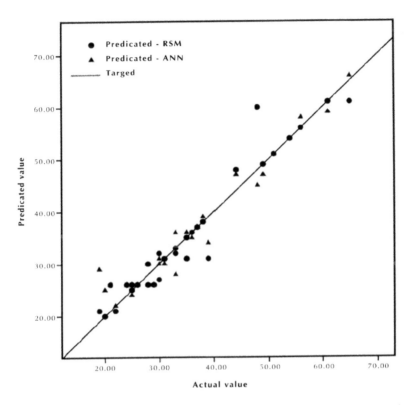

FIGURE 10.48 Comparison between the actual and predicted contact angle of electrospun nanofiber for RSM and ANN model.

TABLE 10.10 Experimental and predicted values by RSM and ANN models

No.	Experimental	Predicted		Absolute error (%)	
		RSM	ANN	RSM	ANN
1	44	47	48	6.41	9.97
2	54	54	54	0.78	0.46
3	61	59	61	3.70	0.42
4	65	66	61	2.06	6.06
5	38	39	38	2.37	0.54
6	49	47	49	5.10	0.68

TABLE 10.10   (*Continued*)

7	51	51	51	0.35	0.45
8	56	58	56	4.32	0.17
9	48	45	60	6.17	24.37
10	30	31	27	4.93	9.35
11	35	36	31	2.34	11.15
12	22	22	21	1.18	4.15
13	30	30	32	1.33	6.04
14	33	28	33	13.94	0.60
15	25	24	25	5.04	0.87
16	26	26	26	0.27	1.33
17	29	26	26	10.10	9.16
18	28	26	26	6.89	5.91
19	25	26	26	4.28	5.38
20	24	26	26	8.63	9.77
21	21	26	26	24.14	25.45
22	31	30	31	2.26	0.57
23	35	31	35	10.34	0.66
24	33	36	32	8.18	2.18
25	37	37	37	0.59	0.34
26	19	29	21	52.11	10.23
27	28	30	30	7.07	8.20
28	39	34	31	12.05	21.30
29	36	35	36	1.72	0.04
30	20	25	20	26.30	2.27
R^2		0.958	0.965		
Mean absolute error (%)				7.83	5.94

APPENDIX D

Variables which potentially can alter the electrospinning process (Figure 10.26) are large. Hence, investigating all of them in the framework of one single research would almost be impossible. However, some of these parameters can be held constant during experimentation. For instance, performing the experiments in a controlled environmental condition, which is concerned in this study, the ambient parameters (i.e. temperature, air pressure, and humidity) are kept unchanged. Solution viscosity is affected by polymer molecular weight, solution concentration, and temperature. For a particular polymer (constant molecular weight) at a fixed temperature, solution concentration would be the only factor influencing the viscosity. In this circumstance, the effect of viscosity could be determined by the solution concentration. Therefore, there would be no need for viscosity to be considered as a separate parameter.

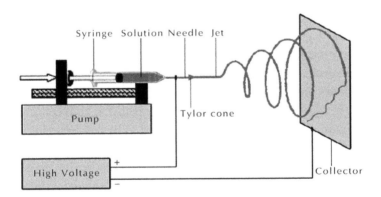

FIGURE 10.49 A typical image of Electrospinning process.

In this regard, solution concentration (C), spinning distance (d), applied voltage (V), and volume flow rate (Q) were selected to be the most influential parameters. The next step is to choose the ranges over which these factors are varied. Process knowledge, which is a combination of practical experience and theoretical understanding, is required to fulfill this step. The aim is here to find an appropriate range for each parameter where dry, bead-free, stable, and continuous fibers without breaking up to droplets are obtained. This goal could be achieved by conducting a set of preliminary experiments while having the previous works in mind along with utilizing the reported relationships.

The relationship between intrinsic viscosity ($[\eta]$) and molecular weight (M) is given by the well-known Mark-Houwink-Sakurada equation as follows:

$$[\eta] = KM^a$$

(10.7)

where K and a are constants for a particular polymer-solvent pair at a given temperature. Polymer chain entanglements in a solution can be expressed in terms of Berry number (B), which is a dimensionless parameter and defined as the product of intrinsic viscosity and polymer concentration ($B = [\eta]C$). For each molecular weight, there is a lower critical concentration at which the polymer solution cannot be electrospun.

As for determining the appropriate range of applied voltage, referring to previous works, it was observed that the changes of voltage lay between 5 and $25kV$ depending on experimental conditions; voltages above $25kV$ were rarely used. Afterwards, a series of experiments were carried out to obtain the desired voltage domain. At $V < 10kV$, the voltage was too low to spin fibers and $10kV \leq V < 15kV$ resulted in formation of fibers and droplets; in addition, electrospinning was impeded at high concentrations. In this regard, $15kV \leq V \leq 25kV$ was selected to be the desired domain for applied voltage.

The use of 5–20cm for spinning distance was reported in the literature. Short distances are suitable for highly evaporative solvents whereas it results in wet coagulated fibers for nonvolatile solvents due to insufficient evaporation time. Afterwards, this was proved by experimental observations and $10cm \leq d \leq 20cm$ was considered as the effective range for spinning distance.

Few researchers have addressed the effect of volume flow rate. Therefore in this case, the attention was focused on experimental observations. At $Q < 0.2ml/hr$, in most cases especially at high polymer concentrations, the fiber formation was hindered due to insufficient supply of solution to the tip of the syringe needle. Whereas, excessive feed of solution at $Q > 0.4ml/hr$ incurred formation of droplets along with fibers. As a result,

$0.2ml/hr \leq Q \leq 0.4ml/hr$ was chosen as the favorable range of flow rate in this study.

Consider a process in which several factors affect a response of the system. In this case, a conventional strategy of experimentation, which is extensively used in practice, is the *one-factor-at-a-time* approach. The major disadvantage of this approach is its failure to consider any possible interaction between the factors, say the failure of one factor to produce the same effect on the response at different levels of another factor. For instance, suppose that two factors A and B affect a response. At one level of A, increasing B causes the response to increase, while at the other level of A, the effect of B totally reverses and the response decreases with increasing B. As interactions exist between electrospinning parameters, this approach may not be an appropriate choice for the case of the present work. The correct strategy to deal with several factors is to use a *full factorial design*. In this method, factors are all varied together; therefore all possible combinations of the levels of the factors are investigated. This approach is very efficient, makes the most use of the experimental data and takes into account the interactions between factors.

It is trivial that in order to draw a line at least two points and for a quadratic curve at least three points are required. Hence, three levels were selected for each parameter in this study so that it would be possible to use quadratic models. These levels were chosen equally spaced. A full factorial experimental design with four factors (solution concentration, spinning distance, applied voltage, and flow rate) each at three levels (3^4 design) were employed resulting in 81 treatment combinations. This design is shown in Figure 10.50.

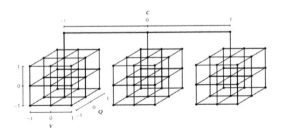

FIGURE 10.50 3^4 full factorial experimental design used in this study.

-1, 0, and 1 are coded variables corresponding to low, intermediate and high levels of each factor respectively. The coded variables (x_j) were calculated using Equation (10.8) from natural variables (ξ_j). The indices 1–4 represent solution concentration, spinning distance, applied voltage, and flow rate respectively. In addition to experimental data, 15 treatments inside the design space were selected as test data and used for evaluation of the models. The natural and coded variables for experimental data (numbers 1–81) as well as test data (numbers 82–96) are listed in Table 3-11 in Appendix.

$$x_j = \frac{\xi_j - \left[\xi_{hj} + \xi_{lj}\right]/2}{\left[\xi_{hj} - \xi_{lj}\right]/2} \qquad (10.8)$$

The mechanism of some scientific phenomena has been well understood and models depicting the physical behavior of the system have been drawn in the form of mathematical relationships. However, there are numerous processes at the moment which have not been sufficiently been understood to permit the theoretical approach. Response surface methodology (RSM) is a combination of mathematical and statistical techniques useful for empirical modeling and analysis of such systems. The application of RSM is in situations where several input variables are potentially influence some performance measure or quality characteristic of the process—often called responses. The relationship between the response (y) and k input variables $(\xi_1, \xi_2, ..., \xi_k)$ could be expressed in terms of mathematical notations as follows:

$$y = f(\xi_1, \xi_2, ..., \xi_k) \qquad (10.9)$$

where the true response function f is unknown. It is often convenient to use coded variables $(x_1, x_2, ..., x_k)$ instead of natural (input) variables. The response function will then be:

$$y = f(x_1, x_2, ..., x_k) \qquad (10.10)$$

Since the form of true response function f is unknown, it must be approximated. Therefore, the successful use of RSM is critically dependent upon the choice of appropriate function to approximate f. Low-order polynomials are widely used as approximating functions. First order (linear) models are unable to capture the interaction between parameters which is a form of curvature in the true response function. Second order (quadratic) models will be likely to perform well in these circumstances. In general, the quadratic model is in the form of:

$$y = \beta_0 + \sum_{j=1}^{k} \beta_j x_j + \sum_{j=1}^{k} \beta_j x_j^2 + \sum_{i<j} \sum_{j=2}^{k} \beta_j x_i x_j + \varepsilon \quad (10.11)$$

where ε is the error term in the model. The use of polynomials of higher order is also possible but infrequent. The βs are a set of unknown coefficients needed to be estimated. In order to do that, the first step is to make some observations on the system being studied. The model in Equation (10.11) may now be written in matrix notations as:

$$\mathbf{y} = \mathbf{X}\hat{\mathbf{a}} + \mathring{a} \quad (10.12)$$

where $\mathbf{y}$ is the vector of observations, $\mathbf{X}$ is the matrix of levels of the variables, $\boldsymbol{\beta}$ is the vector of unknown coefficients, and ε is the vector of random errors. Afterwards, method of least squares, which minimizes the sum of squares of errors, is employed to find the estimators of the coefficients ($\hat{\mathbf{a}}$) through:

$$\hat{\mathbf{a}} = (\mathbf{X}'\mathbf{X})^{-1}\mathbf{X}'\mathbf{y} \quad (10.13)$$

The fitted model will then be written as:

$$\hat{\mathbf{y}} = \mathbf{X}\hat{\mathbf{a}} \quad (10.14)$$

Finally, response surfaces or contour plots are depicted to help visualize the relationship between the response and the variables and see the influence of the parameters. As you might notice, there is a close connection between RSM and linear regression analysis.

After the unknown coefficients (βs) were estimated by least squares method, the quadratic models for the mean fiber diameter (MFD) and standard deviation of fiber diameter (StdFD) in terms of coded variables are written as:

$$MFD = 282.031 + 34.953x_1 + 5.622x_2 - 2.113x_3 + 9.013x_4$$
$$- 11.613x_1^2 - 4.304x_2^2 - 15.500x_3^2$$
$$- 0.414x_4^2 + 12.517x_1x_2 + 4.020x_1x_3 - 0.162x_1x_4 + 20.643x_2x_3 + 0.741x_2x_4 + 0.877_x3x_4$$

$$(10.15)$$

$$StdFD = 36.1574 + 4.5788x_1 - 1.5536x_2 + 6.4012x_3$$
$$- 2.2937x_1^2 - 0.1115x_2^2 - 1.1891x_3^2 + 3.0980x_4^2$$
$$- 0.2088x_1x_2 + 1.0010x_1x_3 + 2.7978x_1x_4 + 0.1649x_2x_3 - 2.4876x_2x_4 + 1.5182x_3x_4$$

$$(10.16)$$

In the next step, a couple of very important hypothesis-testing procedures were carried out to measure the usefulness of the models presented here. First, the test for significance of the model was performed to determine whether there is a subset of variables which contributes significantly in representing the response variations. The appropriate hypotheses are:

$$H_0 : \beta_1 = \beta_2 = \cdots = \beta_k$$
$$H_1 : \beta_j \neq 0 \quad \text{for at least one } j$$

$$(10.17)$$

The F statistics (the result of dividing the factor mean square by the error mean square) of this test along with the p-values (a measure of statistical significance, the smallest level of significance for which the null hypothesis is rejected) for both models are shown in Table 10.11.

TABLE 10.11 Summary of the results from statistical analysis of the models

	F	p-value	R^2	R^2_{adj}	R^2_{pred}
MFD	106.02	0.000	95.74%	94.84%	93.48%
StdFD	42.05	0.000	89.92%	87.78%	84.83%

The p-values of the models are very small (almost zero), therefore it could be concluded that the null hypothesis is rejected in both cases suggesting that there are some significant terms in each model. There are also included in Table 10.11, the values of R^2, R^2_{adj}, and R^2_{pred}. R^2 is a measure for the amount of response variation which is explained by variables and will always increase when a new term is added to the model regardless of whether the inclusion of the additional term is statistically significant or not. R^2_{adj} is the adjusted form of R^2 for the number of terms in the model; therefore it will increase only if the new terms improve the model and decreases if unnecessary terms are added. R^2_{pred} implies how well the model predicts the response for new observations, whereas R^2 and R^2_{adj} indicate how well the model fits the experimental data. The R^2 values demonstrate that 95.74% of MFD and 89.92% of StdFD are explained by the variables. The R^2_{adj} values are 94.84% and 87.78% for MFD and StdFD respectively, which account for the number of terms in the models. Both R^2 and R^2_{adj} values indicate that the models fit the data very well. The slight difference between the values of R^2 and R^2_{adj} suggests that there might be some insignificant terms in the models. Since the R^2_{pred} values are so close to the values of R^2 and R^2_{adj}, models does not appear to be overfit and have very good predictive ability.

The second testing hypothesis is evaluation of individual coefficients, which would be useful for determination of variables in the models. The hypotheses for testing of the significance of any individual coefficient are:

$$H_0 : \beta_j = 0$$
$$H_1 : \beta_j \neq 0 \tag{10.18}$$

The model might be more efficient with inclusion or perhaps exclusion of one or more variables. Therefore the value of each term in the model is evaluated using this test, and then eliminating the statistically insignificant terms, more efficient models could be obtained. The results of this test for the models of MFD and StdFD are summarized in Table 10.12 and Table 10.13 respectively. T statistic in these tables is a measure of the difference between an observed statistic and its hypothesized population value in units of standard error.

TABLE 10.12   The test on individual coefficients for the model of mean fiber diameter (MFD)

Term (coded)	Coeff.	T	p-value
Constant	282.031	102.565	0.000
C	34.953	31.136	0.000
d	5.622	5.008	0.000
V	-2.113	-1.882	0.064
Q	9.013	8.028	0.000
C^2	-11.613	-5.973	0.000
d^2	-4.304	-2.214	0.030
V^2	-15.500	-7.972	0.000
Q^2	-0.414	-0.213	0.832
Cd	12.517	9.104	0.000
CV	4.020	2.924	0.005
CQ	-0.162	-0.118	0.906
dV	20.643	15.015	0.000
dQ	0.741	0.539	0.592
VQ	0.877	0.638	0.526

TABLE 10.13   The test on individual coefficients for the model of standard deviation of fiber diameter (StdFD)

Term (coded)	Coeff.	T	p-value
Constant	36.1574	39.381	0.000
C	4.5788	12.216	0.000
D	-1.5536	-4.145	0.000
V	6.4012	17.078	0.000

TABLE 10.13   *(Continued)*

Q	1.1531	3.076	0.003
C^2	-2.2937	-3.533	0.001
d^2	-0.1115	-0.172	0.864
V^2	-1.1891	-1.832	0.072
Q^2	3.0980	4.772	0.000
Cd	-0.2088	-0.455	0.651
CV	1.0010	2.180	0.033
CQ	2.7978	6.095	0.000
dV	0.1649	0.359	0.721
dQ	-2.4876	-5.419	0.000
VQ	1.5182	3.307	0.002

As depicted, the terms related to Q^2, CQ, dQ, and VQ in the model of MFD and related to d^2, cd, and dv in the model of StdFD have very high p-values, therefore they do not contribute significantly in representing the variation of the corresponding response. Eliminating these terms will enhance the efficiency of the models. The new models are then given by recalculating the unknown coefficients in terms of coded variables in Equations (10.19) and (10.20), and in terms of natural (uncoded) variables in Equations (10.21) and 10.22).

$$MFD = 281.755 + 34.953x_1 + 5.622x_2 - 2.113x_3 + 9.013x_4$$
$$- 11.613x_1^2 - 4.304x_2^2 - 15.500x_3^2 \qquad (10.19)$$
$$+ 12.517x_1x_2 + 4.020x_1x_3 + 20.643x_2x_3$$

$$StdFD = 36.083 + 4.579x_1 - 1.554x_2 + 6.401x_3 + 1.153x_4$$
$$- 2.294x_1^2 - 1.189x_3^2 + 3.098x_4^2 \qquad (10.20)$$
$$+ 1.001x_1x_3 + 2.798x_1x_4 - 2.488x_2x_4 + 1.518x_3x_4$$

$$MFD = 10.3345 + 48.7288C - 22.7420d + 7.9713V + 90.1250Q$$
$$- 2.9033C^2 - 0.1722d^2 - 0.6120V^2 \qquad (10.21)$$
$$+ 1.2517cd + 0.4020CV + 0.8257dV$$

$$StdFD = -1.8823 + 7.5590C + 1.1818d + 1.2709V - 300.3410Q$$
$$- 0.5734C^2 - 0.0476V^2 + 309.7999Q^2 \qquad (10.22)$$
$$+ 0.1001CV + 13.9892CQ - 4.9752dQ + 3.0364VQ$$

The results of the test for significance as well as R^2, R^2_{adj}, and R^2_{pred} for the new models are given in Table 10.14. It is obvious that the p-values for the new models are close to zero indicating the existence of some significant terms in each model. Comparing the results of this table with Table 10.11, the F statistic increased for the new models, indicating the improvement of the models after eliminating the insignificant terms. Despite the slight decrease in R^2, the values of R^2_{adj} and R^2_{pred} increased substantially for the new models. As it was mentioned earlier in the paper, R^2 will always increase with the number of terms in the model. Therefore, the smaller R^2 values were expected for the new models, due to the fewer terms. However, this does not necessarily suggest that the previous models were more efficient. Looking at the tables, R^2_{adj}, which provides a more useful tool for comparing the explanatory power of models with different number of terms, increased after eliminating the unnecessary variables. Hence, the new models have the ability to better explain the experimental data. Due to higher R^2_{pred}, the new models also have higher prediction ability. In other words, eliminating the insignificant terms result in simpler models which not only present the experimental data in superior form, but also are more powerful in predicting new conditions.

TABLE 10.14 ummary of the results from statistical analysis of the models after eliminating the insignificant terms

	F	p-value	R^2	R^2_{adj}	R^2_{pred}
MFD	155.56	0.000	95.69%	95.08%	94.18%
StdFD	55.61	0.000	89.86%	88.25%	86.02%

The test for individual coefficients was performed again for the new models. The results of this test are summarized in Table 10.15 and Table 10.16. This time, as it was anticipated, no terms had higher p-value than expected, which need to be eliminated. Here is another advantage of removing unimportant terms. The values of T statistic increased for the terms already in the models implying that their effects on the response became stronger.

TABLE 10.15 The test on individual coefficients for the model of mean fiber diameter (MFD) after eliminating the insignificant terms

Term (coded)	Coeff.	T	p-value
Constant	281.755	118.973	0.000
C	34.953	31.884	0.000
d	5.622	5.128	0.000
V	-2.113	-1.927	0.058
Q	9.013	8.221	0.000
C^2	-11.613	-6.116	0.000
d^2	-4.304	-2.267	0.026
V^2	-15.500	-8.163	0.000
Cd	12.517	9.323	0.000
CV	4.020	2.994	0.004
dV	20.643	15.375	0.000

TABLE 10.16 The test on individual coefficients for the model of standard deviation of fiber diameter (StdFD) after eliminating the insignificant terms

Term (coded)	Coeff.	T	p-value
Constant	36.083	45.438	0.000
C	4.579	12.456	0.000

TABLE 10.16 (*Continued*)

d	-1.554	-4.226	0.000
V	6.401	17.413	0.000
Q	1.153	3.137	0.003
C^2	-2.294	-3.602	0.001
V^2	-1.189	-1.868	0.066
Q^2	3.098	4.866	0.000
CV	1.001	2.223	0.029
CQ	2.798	6.214	0.000
dQ	-2.488	-5.525	0.000
VQ	1.518	3.372	0.001

After developing the relationship between parameters, the test data were used to investigate the prediction ability of the models. Root mean square errors (RMSE) between the calculated responses (C_j) and real responses (R_j) were determined using equation (D-17) for experimental data as well as test data for the sake of evaluation of both MFD and StdFD models.

$$RMSE = \sqrt{\frac{\sum_{i=1}^{n}(C_i - R_i)^2}{n}}$$ (10.23)

KEYWORDS

- **Biopolymers**
- **Chitin**
- **Chitosan**
- **Electrospinning process**
- **Nanofibers**

REFERENCES

1. Miwa, M., Nakajima, A., Fujishima, A., Hashimoto, K., & Watanabe, T. (2000). *Langmuir, 16*, 5754.
2. Öner, D. & McCarthy, T. J. (2000). *Langmuir, 16*, 7777.
3. Abdelsalam, M. E., Bartlett, P. N., Kelf, T., & Baumberg, J. (2005). *Langmuir, 21*, 1753.
4. Nakajima, A., Hashimoto, K., Watanabe, T., Takai, K., Yamauchi, G., & Fujishima, A. (2000). *Langmuir, 16*, 7044.
5. Zhong, W., Liu, S., Chen, X., Wang, Y., & Yang, W. (2006). *Macromolecules, 39*, 3224.
6. Shams Nateri, A. & Hasanzadeh, M. (2009). *Journal of Computational and Theoretical Nanoscience, 6*, 1542.
7. Kilic, A., Oruc, F., & Demir, A. (2008). *Textile Research Journal, 78*, 532.
8. Reneker, D. H. & Chun, I. (1996). *Nanotechnology, 7*, 216.
9. Shin, Y. M., Hohman, M. M., Brenner, M. P., & Rutledge, G. C. (2001). *Polymer, 42*, 9955.
10. Reneker, D. H., Yarin, A. L., Fong, H., & Koombhongse, S. (2000). *Journal of Applied Physics, 87*, 4531.
11. Zhang, S., Shim, W. S., & Kim, J. (2009). *Materials and Design, 30*, 3659.
12. Yördem, O. S., Papila, M., & Menceloğlu, Y. Z. (2008). *Materials and Design, 29*, 34.
13. Chronakis, I. S. (2005). *Journal of Materials Processing Technology, 167*, 283.
14. Dotti, F., Varesano, A., Montarsolo, A., Aluigi, A., Tonin, C., & Mazzuchetti, G. (2007). *Journal of Industrial Textiles, 37*, 151.
15. Lu, Y., Jiang, H., Tu, K., & Wang, L. (2009). *Acta Biomaterialia, 5*, 1562.
16. Lu, H., Chen, W., Xing, Y., Ying, D., & Jiang, B. (2009). *Journal of Bioactive and Compatible Polymers, 24*, 158.
17. Nisbet, D. R., Forsythe, J. S., Shen, W., Finkelstein, D. I., & Horne, M. K. (2009). *Journal of Biomaterials Applications, 24*, 7.
18. Ma, Z., Kotaki, M., Inai, R., & Ramakrishna, S. (2005). *Journal of Tissue Engineering, 11*, 101.
19. Hong, K. H. (2007). *Polymer Engineering and Science, 47*, 43.
20. Zhang, W. & Pintauro, P. N. (2011). *ChemSusChem, 4*, 1753.
21. Lee, S. & Obendorf, S. K. (2007). *Textile Research Journal, 77*, 696.
22. Myers, R. H., Montgomery, D. C., & Anderson-cook, C. M. (2009). *Response surface methodology: Process and product optimization using designed experiments* (3rd edn.). USA: John Wiley and Sons.
23. Gu, S. Y., Ren, J., & Vancso, G. J. (2005). *European Polymer Journal, 41*, 2559.
24. Dev, V. R. G., Venugopal, J. R., Senthilkumar, M., Gupta, D., & Ramakrishna, S. (2009). *Journal of Applied Polymer Science, 113*, 3397.
25. Galushkin, A. I. (2007). *Neural networks theory.* Moscow Institute of Physics & Technology: Springer.
26. Ma, M., Mao, Y., Gupta, M., Gleason, K. K., & Rutledge, G. C. (2005). *Macromolecules, 38*, 9742.
27. Haghi, A. K. & Akbari M. (2007). *Physica Status Solidi A, 204*, 1830.

28. Ziabari, M., Mottaghitalab, V., & Haghi, A. K. (2009). *Nanofibers: Fabrication, performance and applications.* USA: W. N. Chang, Nova Science Publishers.
29. Ramakrishna, S., Fujihara, K., Teo, W. E., Lim, T. C., & Ma, Z. (2005). *An introduction to electrospinning and nanofibers.* Singapore: World Scientific Publishing.
30. Zhang, S., Shim, W. S., & Kim, J. (2009). *Materials and Design, 30,* 3659.
31. Kasiri, M. B., Aleboyeh, H., & Aleboyeh, A. (2008). *Environmental Science and Technology, 42,* 7970.

INDEX

A

Acetate fibers, 324
Akhmetshina, L. F., 53–112
Aleksanyan, K. V., 13–33
Aloev, V. Z., 43–52
Alternaria, 307, 319, 328
Amino acids, 336
Analysis of variance (ANOVA), 356, 357–360
Artificial fibers biodamaging, 323–324
Artificial neural network (ANN), 353, 356–357, 360–361
Aspargillus niger, 319
Aspergillus, 307, 314
Aspergillus flavus, 314
Aspergillus niger, 314
Avogadro number, 39
Avrami's equation, 99

B

B. megatherium, 328
B. mesentericus, 328
B. mycoides, 328
B. subtilis, 328
Bacillus, 307, 314
Bacillus agri, 328
Bacillus subtilis, 314
Bacterial plasmids
    Escherichia coli, 115
    GACGTT-motif, 114
    materials
        CpG motifs, 115
        GTCGTT motif, 120
        isolation and purification, 121
        Luria-Bertani (LB) agar plates, 116
        MinElute Gel Extraction Kit, 116
        NdeI and EcoRI, 117
    phosphorylation, 116
    oligodeoxynucleotides (ODNs), 114
    phosphorothioate linkages, 115
    protein and DNA vaccines, 114
    synthetic CpG ODNs, 114
    unmethylated CpG dinucleotides, 114
Basis superpositional error (BSSE), 75
Bast fiber biodamading, 320–322
Biodegradable compositions
    binary compositions, 16, 23
    chemical method, 18
    crystallinity degree, 20
    DSC curves, 19
    elastic modulus, 23–26
    electron micrographs, 22
    extrusion mixing, 15
    fungus resistance, 28
    grinding process, 15
    iodine reaction, 18
    LDPE melting parameters, 20
    low density polyethylene (LDPE), 15–16
    mechanical mixing, 19
    mechanical parameters, 25
    mechanical properties, 15
    mechanical tests, 23
    melting curves, 20–21
    melting enthalpies, 19
    micrographs, 29

particle size distribution histograms, 17
photodegradation, 26
photooxidation, 26
polyethylene, 26
poly(ethylene oxide) (PEO), 15–16
polymer materials, 14
polymer mixing, 15
polymer system stability, 26
polypropylene, 26
polysaccharide, 14, 16
rotor disperser, 16
scanning electron microscopy, 21
shear deformations, 15–16
starch-LDPE-chitin composition,
30–31
synthetic polymers, 25
ternary composition, 23
ternary compositions, 16
TGA curves, 26–28
thermal stability, 26
thermogravimetric analysis, 26, 28
traditional synthetic polymers, 14
ultimate tensile strength, 23–26
weight loss curves, 30
X-ray diffraction, 18
X-ray pattern, 18–19

C

Carbon nanocomposites influence mechanisms
    band intensity, 93
    changes in IR spectrum, 92
    chemical bonds, 88
    composition hardening process, 86
    distribution of, 87–88
    epoxy compounds, 91
    IR spectrum of, 89–90
    isomethyltetrahydrophthalic anhy-
    dride, 91
    laser analyzer, 86
    liquid phase, 85

    nanoparticle distribution, 91
    nanoparticle oscillations, 90
    optic density, 93
    particle sizes, 91
    phenolformaldehyde, 91
    phenolrubber polymers alcohol, 91
    physical-mechanical characteris-
    tics, 91
    polyethylene polyamine, 91
    polyethylene polyamine IR spec-
    trum, 89
    polymethyl methacrylate dichloro-
    ethane, 91
    polyvinyl chloride compositions, 91
    skeleton bonds oscillations, 90
    sodium lignosulfonate, 92–93
    suspension optical density, 86
    toluene, 91
    water-alcohol suspensions, 86
    X-ray photoelectron, 86
Cell membrane complex (CMC), 326
Central composite design (CCD), 355
Chaetomium, 314
Chaetomium globosum, 314, 319
Chashkin, M. A., 53–112
Chemical fibers, 306–307
Cladosporium, 307, 319
CNTs nano-composites
    chitosan porous membranes, 248
    covalent grafting
        acetic acid, 250
        acyl chloride, 250
        anhydrous dimethyl for-
    mamide, 250
        carboxylic acid groups, 250
        lactic acid, 250
        potassium persulfate, 250
        thionyl chloride, 250
    crosslinking-casting-evaporation
        acetylcholinesterase, 248
        amperometric acetylthiocho-
    line sensor, 248
    electrical properties, 248

electro-deposition method
cathode surface, 249
glucose biosensor, 249
electrospinning, 252–253
electrostatic interaction, 250–251
freeze-drying, 252
layer-by-layer self assembly, 252
mechanical properties, 247
microwave irradiation, 251
solution-casting-evaporation
amperometric hydrogen peroxide biosensor, 247
amperometric laccase biosensor, 247
chitosan film, 247
electrochemical biosensors, 246
glassy carbon electrode, 246
glucose biosensors, 246
surface deposition cross-linking, 249
wet-spinning, 252
Compositions modification processes, carbon nanocomposites
epoxy resins modification
amine groups, 101
coordination bonds, 101
metal ions, 101
polyethylene polyamine IR spectra, 100
polymeric composite materials, 100
quantum-chemical modeling methods, 101
semi-empirical methods, 101
thermal stability, 100, 102
ultrasound processing, 101
foam concrete modification
acetate groups, 97
Avrami's equation, 99
band intensity, 99
breaking stresses, 98

hydroxyl groups, 97
oscillation energy, 99
polyvinyl alcohol, 97
glues modification, 102
alcohol suspension, 103
ammonium phosphates, 103
ammonium polyphosphate, 103
combustion process, 103
flammability test, 104
nanostructure distribution, 103
organic solvent, 103
synthetic resin, 103
synthetic rubber, 103
polycarbonate modification
ethylene dichloride, 105
IR spectra of, 109
linear structures, 108
optical density curves, 106–107
polymethyl methacrylate, 105
surface energy, 104
thermal conductivity, 109
thermal-physical characteristics, 105, 110
thermal-physical methods, 106
polymeric materials modification
acetone, 94
benzene, 94
coagulation process, 95
dichlorethane, 94
dynamic viscosity, 96
ethanol, 94
isomethyl tetra hydrophtalic anhydrite, 94
kinematic viscosity, 96
methylene chloride, 94
microphotographs, 95–96
oleic acid, 94
optical density, 94
polyethylene polyamine, 94
suspension stability, 95
toluene, 94

Cotton fiber, 309–310, 312
Cyclic diguanylate enzymatic synthesis
    allosteric sites, 126
    diguanylate cyclase (DGC), 126
    enzymatic activity, 130
    gluconacetobacter xylinum, 127
    guanosine-5'-triphosphate (GTP)
    molecules, 127
    high thermostability, 130
    liquid chromatography, 133
    modified Biuret method, 129
    phosphatase activity, 131
    sodium dodecyl sulfate (SDS), 129
    staphylococcus aureus, 126
    Thermotoga maritima, 126
Cytophaga, 307

D

Dibirova, K. S., 35–42
Dicarboxylic acids, 308
Differencial scanning calorimetry (DSC), 2
Diguanylate cyclase (DGC), 126
Dynamic light scattering (DLS), 279

E

Einstein's formula, 79
Electrospinning, processing parameters,
    254
Electrospun fiber mat, 365–366
Electrospun fibers, 351–352
Electrostatic repulsive force, 261, 263
Enzymes, 309
Euclidean space, 38, 46

F

Fisher's statistical test, 356
Fourier transform infrared spectra, 278
Fusarium, 307

G

Green nanofibers
    acetic acid, 275, 281
    ambient conditions

        atmosphere pressure, 274
        humidity, 274
        temperature, 274
    ambient parameters, 273
    amide functional group, 280
    analysis of variance, 306
    applied voltage, 285, 306
    biomedical applications, 273
    biopolymers, 274, 279
    carbon nanotube dispersion, 305
    carbon nanotubes (CNT), 275
    charge density, 275
    chitin, 273
        chemical structures of, 275
    CHT-MWNTS dispersions
        organic acid, 276
        protocols, 277
    dichloromethane (DCM), 275
    dynamic light scattering technique,
    278–279
    effect of CHT concentration, 284
    electrical conductivity, 285
    electrical properties, 275
    electrospinning process, 274
        positive electrode, 277
        temperature, 305
        tip-to-collector distance, 277
    electrospun fiber, 281, 285
    fiber diameter, 285, 306
    formic acid, 281
    Fourier transform infrared spectra,
    278, 281
    hydrodynamic diameter, 279
    hydrogen bonds, 281
    hydroxyl groups, 276
    inorganic acids, 275
    mechanical properties, 275
    multiwalled nanotube (MWNT),
    275
    nanofiber thin film, 279
    natural materials, 273
    organic acids, 275

polyethylene oxide, 275
polymeric chains, 275
polymer solution, 273
processing conditions
    applied voltage, 274
    needle diameter, 274
    spinning distance, 274
    volume flow rate, 274
scanning electronic micrographs, 282–283, 287
scanning electron microscope, 278
single walled nanotube (SWNT), 275
solution concentration, 305
solution properties, 273
sonication energy, 280
surface tension, 275
tip-to-collector distance, 286, 305
transmission mode, 278
triflouroacetic acid (TFA), 275

H

Haghi, A. K., 139–193, 195–270, 271–382
Hamrang, A., 139–193, 195–270
Hartree-Fock method, 55
Hasanzadeh, M., 271–382
Howell, B. A., 35–42, 43–52
Hydrogen bonds, 312
Hydroxyl groups, 311–312

I

IR-spectroscopy method, 337

K

Khokhriakov, N. V., 53–112
Kodolov, V. I., 53–112
Kolmogorov-Avrami equation, 45
Korovashkina, A. S., 113–123, 125–135
Kozlov, G. V., 35–42, 43–52
Kvach, M. V., 125–135
Kvach, S. V., 113–123, 125–135

L

Linear low density polyethylene (LLDPE), 37
Linear regression analysis, 373
Low density polyethylene (LDPE), 15–16

M

Magomedov, G. M., 35–42
Mark-Houwink-Sakurada equation, 370
Maxwell's theory, 245
Mean fiber diameter (MFD), 374
Micrococcus, 307
Moghadam, B. H., 271–382
Mottaghitalab, V., 271–382
Mucor, 314
Multiwalled nanotube (MWNT), 275

N

Nanocomposites polyethylene reinforcement mechanism
    ammonium ions, 37
    Avogadro number, 39
    cluster model parameters, 40
    crystallinity degree, 37
    elasticity moduli, 36
    Euclidean space, 38
    linear low density polyethylene, 37
    macromolecule cross-sectional area, 38
    matrix polymer structure, 36
    mechanical tests, 38
    modified Na+-montmorillonite (MMT), 37
    molecular weight, 39
    nanoclusters dimensional effect, 37
    nanocomposite elasticity modulus, 40
    nanofiller density, 40
    Poisson's ratio, 38
    polymer chain statistical flexibility indicator, 38
    polymer composites, 36

polymer density, 39
reinforcement percolation model, 40
relative volume fractions, 40
tensile tests, 37
Nanocomposites polypropylene crystallization kinetics
    carbon fibers, 48
    carbon nanotubes (CNT), 45
    components mixing, 45
    crystallinity degree, 51
    differential scanning calorimetry method, 45
    elasticity moduli, 50
    Euclidean space, 46
    fractal model, 44
    high density polyethylene (HDPE), 44
    hydrocarbons chemical deposition, 45
    Kolmogorov-Avrami equation, 45
    Kolmogorov-Avrami exponent, 48–49
    molecular mobility level, 46
    nanocomposite matrix crystallization process, 47
    nanocomposite structure fractal dimension, 46
    nanofiller mass content, 48
    nanofiller particles, 44
    nanofiller volume contents, 47
    polydispersity index, 44
    polymer chain statistical flexibility indicator, 46
    polymer matrix crystallinity degree, 44
    polymers structure, 44
    yield stress, 46
Nanofibers
    applied voltage, 261–263
    concentration, 258–260
    SEM images, 255–255

Nanomaterial
    analysis of variance (ANOVA)
        electrospun fiber mat, 220–221
        regression analysis, 222
    applications
        biopolymers, 245
        carbon nanotube composite, 244
        food packaging, 245
        optical properties, 244
        photovoltaic properties, 244
        sustainable environment, 244
    carbon nanotube nanofluids
        Maxwell's theory, 245
        non Newtonian behavior, 245
    characterization
        angle measurement set up, 212
    contact angle
        average fiber diameter (AFD), 210
        superhydrophobic surface, 210
    electric field, 200
    electrospinning
        process, 199
        room temperature, 211
        syringe pump, 211
    electrospinning set up, 199
        drum collector, 201
        screen collector, 201
    electrospun fiber mat
        interfibrillar distance, 231
        nanoscale fibers, 231
    electrospun nanofibers, 202
    experimental design
        central composite design, 212
        response surface methodology, 212
    materials
        dimethylformamide, 211
    molecular weight
        rheological and electrical properties, 204

nanocomposites mechanical and electrical properties
    natural polymer, 242
    neat biopolymers, 243
nanocomposites production
    algal origin, 233
    amphiphilic molecules, 238
    animal origin, 233
    anionic surfactants, 238
    biomaterials, 233
    biopolymer, 232–233
    carbon nanotubes (CNTs), 236
    cationic surfactants, 238
    chiotosan applications, 235
    chitin structure, 234
    chitosan matrix, 236
    deacetylation degree (DD), 234
    degree of graphitization, 237
    microbial origin, 233
    multi-walled carbon nanotubes (MWNTs), 236
    nanobiocomposites, 232
    natural and synthetic biopolymers, 233
    non ionic surfactants, 238
    plant origin, 233
    polymerization processing, 237
    polysaccharides, 233
    single-walled carbon nanotubes (SWNTs), 236
    thermal and electrical conductivity, 237
nanofibers morphological analysis
    chain mobility, 214
    collector distance, 217
    pan solution, 213
    SEM image, 215
    spinning distance, 216
    volume flow rate, 218
nanotube composites
    bulk materials, 239
    carbon black (CB), 241
    crystalline orientation, 240
    crystallinity, 240
    double-walled carbon nanotube (DWNT), 241
    entanglement, 240
    interfacial stress transfer, 239
    straightness, 240
optimal conditions
    desirability, 230
    spinning distance, 230
processing condition
    applied voltage, 207
    electrical force, 207
    electrostatic repulsive force, 208
    feed rate, 208
    morphology, 208
    solution concentration, 209
response surfaces for AFD
    applied voltage, 226
    solution concentration, 225
    tip to collector distance, 226
    volume flow rate, 226–227
response surfaces for CA
    applied voltage, 228
    solution concentration, 228
    tip to collector distance, 229
    volume flow rate, 229
solution concentration
    beads and fibers, 203
    visco-elastic force, 204
solution conductivity, 207
solution properties, 202
surface tension
    elastic skin, 205
    electrospun nanofiber, 206
    upper and lower boundaries, 205
viscosity
    electrospun nanofibers, 203
    fiber diameter, 203
voltage supplier, 200

Nanopolymers, 166
   cell membrane, 186
   condition processing
      applied voltage, 158
      feed rate, 159
   control morphology, 162
   copolymers
      addition polymers, 150
      alternating, 149
      block, 149
      condensation polymers, 150
      graft, 149
      random, 149
      unsaturated monomers, 151
   Coulomb's law, 160
   drawing
      micromanipulator, 143
      micropipette, 143
   ECG system, 188
   electrical current, 185
   electrohydrodynamic instability theory, 160
   electroless plating, 191
   electroless printing, 191
   electrospinning
      electrode collector, 146
      electrostatic spinning, 145
      high-voltage power supply, 146
      polymer solution, 146
      spinneret (needle), 146
      Taylor cone, 146
   electrospinning method, 163
   fabric sensor, 190
   health deterioration, 189
   heart cells, 186
   homopolymer, 149
   hydrodynamics
      dispersion laws, 167
      electric field, 167
      fluid dynamics, 165
   intelligent textiles, 190

   lattice Boltzmann method, 164
   low basis weight, 183
   mass production, 184
   mathematical and theoretical modeling, 161
   models
      ac-electrospinning mathematical model, 177
      coupled field forces mathematical model, 173
      electrospinning, 178
      leaky dielectric model, 170
      magnetic electrospinning mathematical model, 179
      multiple jet model, 179
      nanoporous materials model, 180
      non linear model, 173
      shape evaluation model, 172
      slender-body model, 174
      viscoelastic fluids model, 176
      whipping model, 171
   momentum equation, 160
   monitoring system, 187
   morphology and diameter
      electric field, 182
      empirical model, 182–183
   nanofibers
      dry spinning, 143
      gel-state spinning, 143
      melt spinning, 143
      viscose, 142
      wet spinning, 143
   natural polymers
      polysaccharides, 154
      proteins, 154
      starch polymer, 154
   non newtonian fluid mechanics method, 168
   phase separation
      freeze-drying, 145
   physiological monitoring, 187–188

proteins
amino acids, 155
radial momentum balance, 161
self-assembly, 145
semi- empirical methods, 184
slender-body theory, 159
slender jet model, 163
synthetic polymers
fibers, 153
plastics, 154
plastics and rubbers, 153
template synthesis
nanoporous membrane, 144
semiconductors, 145
tubules and fibrils, 145
X-ray detection
flow rate and viscosity, 169
radiographic films, 169
radiographic method, 169
shield film, 169
Natural fibers, 306
Nuclear magnetic resonance (NMR), 2

P

Penicillium, 307, 314
Pershin, Y. U., 53–112
Photonic crystals, 264
Planck constant, 79
Poisson's ratio, 38
Polyamide fibers, structure and properties,
341–345
Polyethylene oxide (PEO), 275
Polyethylene polyamine (PEPA), 77
Polymer concentration, 253
Polymeric composite materials (PCM),
100
Polyotov, Y. A., 53–112
Polyvinyl alcohol (PVA), 275
Potato starch (PS), 3
Proteus vulgaris, 328
Prut, E. V., 13–33
Pseudomonas, 307, 314
Pseudomonas aeroginosa, 328
Pullularia, 319

Q

Quadratic model, 373
Quantum chemical investigation
basis sets of complex molecules
(BSSE), 65
basis superpositional error, 75
binding energy, 66, 69
bond energy, 55, 63
bond lengths, 73–74
carbon envelope
heptagonal defect, 72
hexagonal defect, 71
carbon nanoparticles, 62
chemical bonds, 63
complex geometrical parameters,
59
copper clusters, 75
electron correlation, 57
electron density, 74
energy models, 55
energy parameters analysis, 57, 66
equilibrium geometric structure, 56,
58, 64, 67
ethyl alcohol molecule, 55, 60
fullerene molecule structure, 57
geometric parameters analysis, 74
Hartree-Fock method, 55, 57, 65
hydrogen bond, 60, 62
hydrogen bond energy, 73
hydroxyfullerene, 57, 62, 69
hydroxyl group, 74
interaction energy, 55, 63, 64,
75–76
metal-carbon nanocomposites and
polymeric materials
ab initio method, 78, 83–84
acetiylacetone, 77
acetylacetone different states,
78
acetylacetone IR spectra,
81–82
curing process, 77

diacetonealcohol, 77
Einstein's formula, 79
epoxy resin, 77
ligand shell, 77
phase distribution, 77
Planck constant, 79
PM3 method, 85
polyethylene polyamine, 77
semi-empirical method, 78, 79
silver filler, 77
silver powder, 78
transition energy, 79
vibrational analysis, 78, 79
model system, 61, 70
molecular system, 59, 67–68
molecule energy, 55
nanoparticle model, 71
α-naphthol molecule, 63
β-naphthol molecules, 63
oxygen atoms, 72
pentagonal defect, 74
perturbation theory, 57
scaling factor, 66
semi-empirical method, 55, 69
transition metals, 63
water molecule, 55, 61
water structure, 69
zero-point vibration energy
(ZPVE), 65

R

Response surface methodology (RSM),
289, 353–356
Rhizopus nigricans, 314
Rogovina, S. Z., 13–33
Root mean square errors (RMSE), 380
Rymko, A. N., 125–135

S

Sarcina, 307
Scanning electron microscopy (SEM), 21
Sergeev, A. I., 1–12

Shilkina, N. G., 1–12
Silk fibers
amino acid, 287
cellulose tubular membrane, 290
effect of
applied voltage, 296
diameter distribution, 296
electric field, 292
electrospun nanofibers, 291
fiber diameter, 292
fiber distribution, 293–297
formic acid, 297–298
mean fiber diameter, 298
polymer solution, 297
SEM micrograph, 293–297
solution viscosity, 297
electric field effect
applied voltage, 298
electrospinning conditions, 299
mean fiber diameter, 298, 299
electrospinning, 290–291
parameters, 289
process, 287
electrospinning temperature effect
atmospheric pressure, 299
electrical forces, 300
electrospinning process, 300
electrospun nanofibers, 299
electrospun ribbons, 299
fiber diameter, 301
SEM micrograph, 300–301
surface tension, 300
fibroin
alanine, 289
glycine, 289
serine, 289
formic acid, 289
hexafluoroacetone, 289
hexafluoro-2-propanol, 289
model
applied voltage, 302

effect of electrospinning parameters, 304–305
electrospinning parameters, 302
fiber diameter, 302
P-value, 303–304
regression analysis, 303
response surface methodology, 302
second-order model, 302
response surface methodology, 289
spinning solution, 290
tensile strength, 288
Single walled nanotube (SWNT), 275
Small angle neutron scattering (SANS), 2
Standard deviation of fiber diameter (StdFD), 374, 379
Starches, 1
amylose content, 5
binary systems, 2
biodegradable compositions, 13–21
biopolymers, 4
convectional heating, 6
differencial scanning calorimetry, 2
additional unfrozen water (AUW), 3
aluminum pans, 3
frozen water (FW), 3
starch-water systems, 3
unfrozen water (UW), 3
gelatinization, 5, 10
gel structure, 5
granule structure, 6
hydrogen bonds, 5
hydrogen bounds, 9
inorganic acid, 2
inter-granular water, 9

less mobile proton population, 7–8, 10
microwave irradiation, 2, 6
NMR-relaxation method, 2
Care-Purcell-Meiboom-Gill (CPMG) pulse sequence, 4
relaxation curves, 4
spin-spin relaxation times, 4
proton transverse relaxation curves, 6
silicic acid, 6
small angle neutron scattering, 2
spin-spin relaxation times, 7–8, 10
starch-water interaction, 5
starch-water mixture, 4
water fraction, 6
water molecules, 4
water populations, 6–7
Staroszczyk, H., 1–12
Synthetic fibers biodamages, 338–341

T

Taylor cone, 146
Textile fibers, 306
Textile material protection, 345–349
Trichoderma, 307, 314
Trichothecium roseum, 314
Trineeva, V. V., 53–112

W

Wasserman, L. A., 1–12
Wool fiber biodamaging, 324–329

Z

Zaikov, G. E., 35–42, 43–52
Zhirikova, Z. M., 43–52
Zinchenko, A. I., 113–123, 125–135